Satellites of the Outer Planets

Satellites of the Outer Planets

WORLDS IN THEIR OWN RIGHT

Second Edition

David A. Rothery

New York   Oxford
Oxford University Press
1999

Oxford University Press

Oxford   New York
Athens   Auckland   Bangkok   Bogotá   Buenos Aires   Calcutta
Cape Town   Chennai   Dar es Salaam   Delhi   Florence   Hong Kong   Istanbul
Karachi   Kuala Lumpur   Madrid   Melbourne   Mexico City   Mumbai
Nairobi   Paris   São Paulo   Singapore   Taipei   Tokyo   Toronto   Warsaw

and associated companies in
Berlin   Ibadan

Published by Oxford University Press, Inc.
198 Madison Avenue, New York, New York 10016

Oxford is a registered trademark of Oxford University Press

Library of Congress Cataloging-in-Publication Data
Rothery, David A.
Satellites of the outer planets: worlds in their own right /
David A. Rothery.—2nd ed.
p.   cm.
Includes bibliographical references and index.
ISBN 0-19-512555-X
1. Outer planets—Satellites.   I. Title.
QB401.R67   1999
523.9'8—dc21   98-52813

9 8 7 6 5 4 3 2 1

Printed in Hong Kong
on acid-free paper

To Edwin and Elanor

Preface to First Edition

Time was when, very properly, geologists confined their attention to the Earth. After all, its profusion of rivers and mountains, beaches and cliffs, and above all the fact that we live there are what enabled the science of geology to get going. Geology is based upon observations made on the Earth. Concepts such as the principle of superposition (younger deposits tend to overlie older deposits) and uniformitarianism (the present is the key to the past) are all "homegrown."

During the latter part of this century, the Earth sciences were revolutionized by the development of techniques and the discovery of processes that give us insights into how the Earth works as a whole. Plate tectonics is the obvious example of this. At the same time, however, the data were beginning to accumulate, at first very slowly but then increasing rapidly, that mean we no longer have just the one planet to explore and to seek to understand. Serious geological studies of the Moon began in the early 1960s, first by telescopic observations, then using high-resolution cameras in orbit round the Moon, and soon by manned landings and the return of samples. Space probes to other planets, notably Mars, began to reveal evidence of volcanism, faulting, erosion, and transport by wind and water that were at once familiar and yet at the same time subtly different from what we are used to at home.

The other terrestrial planets, Mercury, Venus, the Moon, and Mars, so called because they resemble the Earth in having a rocky outer layer upon which their geological histories are written, have received a fair amount of attention. There are now several good texts on comparative planetology that deal with these worlds in detail, although there has been a curious and unfortunate reluctance within much of the Earth science community to read them.

In this book I want to take a look at a different class of bodies, the moons of the outer planets: Jupiter, Saturn, Uranus, and Neptune. Most of these lack a rocky crust but are nonetheless surfaced by hard, rigid material—ice of various kinds—that behaves in almost all respects like rock. They constitute about a quarter of the mappable terrain in the solar system, only slightly less then the total surface area of the Earth (including the ocean floor). These worlds are important objects for study by geologists if they wish to gain the most that comparative planetology can offer in the way of insights into how the Earth and other planet-sized bodies have evolved and what processes may occur on and beneath their surfaces.

To the astronomer, the moons of the outer planets used to be convenient points of light: markers in the game of celestial mechanics useful for measuring the mass of the planet around which they orbit, or distant clocks for determining the speed of light. Now, thanks to the outstanding achievements of the two *Voyager* space probes, we have been privileged to see their surfaces from close by and can recognize them for what they truly are: worlds in their own right upon which the forces of geology have left their marks in a variety of similar and yet intruigingly varied ways.

This book is not intended as an all-embracing treatise on the satellites of the outer planets. There are other works, noted in the bibliography, that come much closer to fulfilling that role. Here, the smaller satellites are discussed only in passing, and the larger ones are treated from a geological point of view. The matter of the origin of these bodies is covered only briefly, although I believe in sufficient depth for their significance in understanding the formation and evolution of the solar system to be appreciated.

As is standard practice these days, the term "billion" is used to mean a thousand million (10^9) and not a million million (10^{12}). I have attempted to avoid jargon as much as possible, so as to make this book accessible to a wide audience. I have tried to explain terms that may be unfamiliar (or used in an unfamiliar sense) where they first appear, and have included a glossary at the end.

This book would not have been possible without the generous provision of illustrative material by several people, especially Alfred McEwen and Marion Rudnyk, or the efforts of my friends and colleagues who read parts of it or took time to discuss various issues with me, especially Helen Rothery, and John Spencer who read the whole thing in draft form. Nor could it have been written without the U.S. taxpayer, whose government has funded the lion's share of mankind's exploration of the outer solar system. To all of these people, my thanks. The specific sources of the illustrative material used are listed in the figure and plate credits, on p. 229–30.

Pury End   D.A.R.
September 1991

Preface to Second Edition

My fascination with the geology of the outer solar system has grown with the years. To judge by the first edition's favorable reviews, this is a feeling shared by many others. This new edition is revised and re-illustrated in the light of discoveries made by the space probe *Galileo* in both its primary mission and its extended *Galileo Europa Mission*. These have provided amazingly detailed pictures of Jupiter's icy satellites, refined our knowledge of their surface compositions and internal structures, and documented the changing pattern of active volcanism on Io. Europa, especially, has proved to be a delight.

No space probe has ventured beyond Jupiter since *Voyager-2* (the star of the first edition), and our next close-up study of the Saturn system will not begin until November 2004 when *Cassini* arrives. However, images from the Hubble Space Telescope, ground-based spectroscopic work, and continued interpretation of *Voyager* images have advanced our appreciation of the outer solar system beyond Jupiter in many ways of which I have tried to take account. Notable among these are our new knowledge of Pluto and the Kuiper belt, and our first inklings of the surface patterns on Titan.

Once again, I express my gratitude to the friends and colleagues who have been willing to share their thoughts with me, and in particular to John Spencer, who commented on many of the more heavily revised sections.

Silverstone   D.A.R.
March 1999

Contents

Satellites of the Outer Planets

1 Introduction

So many worlds, so much to do,
So little done, such things to be
Tennyson,
In Memoriam

Galileo made a remarkable discovery during the winter of 1610 when he turned one of the earliest telescopes on to the planet Jupiter and saw that four smaller bodies were moving round it. Many philosophers of the time refused to accept the truth of this observation, because it undermined the long-held belief that Earth was at the center of the universe and that all the heavenly bodies (the Sun included) must go around it.

For a while, Jupiter's galilean satellites, as they became known, played a major role in the advance of science through the information they gave about the universe around us. As Galileo realized, the fact that they do indeed go around the planet Jupiter was a strong argument in favor of the Copernican (Sun-centered) view of the cosmos and against the then established view, which had Earth at the center of everything. Sixty years or so later, after the periodic motions of these satellites had become sufficiently documented, the Danish astronomer Ole Rømer measured slight differences between the predicted and observed times at which each of these moons disappeared into Jupiter's shadow. He correctly realized that the orbits are as regular as clockwork, but because the distance between Earth and Jupiter varies, so does the time taken for the light to reach us. Thus Rømer was able to show that the speed of light must be finite, and in fact made a surprisingly accurate determination of its value. Soon, with the benefits of Newton's gravitational theory and laws of motion, the size and period of the orbits of Jupiter's satellites enabled the enormous mass of the planet itself to be calculated.

Once a family of satellites had been documented around Jupiter, it was comparatively unremarkable to discover that the other outer planets in the solar system also have moons. By 1700, five satellites of Saturn were known. Within six years of his discovery of the planet Uranus in 1781, Sir William Herschel had found two of its moons, and the largest moon of Neptune was discovered less than three weeks after the planet itself in 1851. Now and again, smaller and fainter moons were found, and by 1950 the tally of outer planet satellites ran Jupiter, eleven; Saturn, nine; Uranus, five; and Neptune, two.

By this time such bodies had long been little more than astronomical curios. What was known about them other than the mechanics of their orbits (which were already known in great detail) could probably have been written in the space of this book so far. However, about this time people did begin to think seriously about what these bodies might be made of. Spectroscopic observations began to reveal a few curiosities about the composition of their surfaces and, in the case of Titan (the largest moon of Saturn), about its atmosphere. Even so, these satellites remained specks of light to be observed using large telescopes. No one could say what any of them would look like close-up; no one knew what forces had shaped their surfaces or what their internal structures were. All this changed in the dramatic ten and a half years between the encounters with the moons of Jupiter by the space probe *Voyager-1* in March 1979 and the flyby of the Neptune system by *Voyager-2* in August 1989.

When geologists pay any attention to the geology of other worlds (which is usually not often enough), they think mostly of the other terrestrial planets: Mars, with its boulder-strewn plains, giant extinct volcanoes, and dry canyons; Venus, with its greenhouse atmosphere and volcanic surface; Mercury, with its impact-scarred surface bearing also the marks of a history of global compression; or our own Moon, with its impact craters and basalt lava plains. However, the *Voyager* missions showed other families of worlds, planet-sized in their dimensions, which have geological histories that are equally fascinating, if not more so (fig. 1.1). Thanks to *Voyager* and the orbital tour of the Jupiter system by the *Galileo* mission (1995–99), many of the satellites of the outer planets are now revealed as tangible worlds with a greater range of geological processes on display than almost anyone would have dared hope for. It is on these, the larger among the moons of Jupiter, Saturn, Uranus, and Neptune, that I concentrate in this book. These are the worlds in their own right whose story I wish to tell.

1.1 GEOLOGY ON PLANETARY SATELLITES

It is possible to divide the satellites of the outer planets into several groups. Among the larger bodies—those with radii in excess of about 200 km—there is a continuum of types ranging from currently active worlds with continually erupting volcanoes, through worlds that have been volcanically and tectonically active recently (geologically speaking, at least), to those that have been inactive for so long that all traces of geological processes have been obliterated by impact craters. Some basic facts about these worlds are presented in table 1.1, and the satellite families of the outer planets are shown to scale in figure 1.2. Two of these satellites are bigger than the planet Mercury (although being of lower density their masses are less), these and two others are both bigger and more massive than Earth's Moon, and a total of six are bigger and more massive than the planet Pluto.

As figure 1.2 shows, in addition to the major satellites there is a plethora of smaller bodies, ranging in size from about 200 km in radius down to the smallest detectable during the *Voyager* flybys (about 20 km across). These are generally irregular in shape and represent collisional fragments of larger moons or captured rocky or icy asteroids and comet nuclei; they are too small for their own gravity to have pulled them into the near-spherical (strictly speaking, spheroidal) shapes of the larger satellites (fig. 1.3). The satellite system of each giant planet begins closest to the planet with the ring system (consisting of dust to house-size debris) and associated moonlets; then come larger satellites in nearly circular orbits close to the planet's equatorial plane, followed by an outer family of small satellites in irregular (elongated, inclined, and sometimes retrograde) orbits. The inner moonlets and outer irregular satellites would make a fascinating study, but we know comparatively little about most of them, other than that most have densities less than that of ice and so are probably heaps of rubble, and that the inner moonlets are probably the source of the ring material, which is blasted off them as a result of impacts by interplanetary meteoroids. From the point of view of this book, they are not really "worlds" in that they never had histories that can be described in terms of geological processes. These are generally passed over in what follows in order to concentrate on the larger bodies.

Most of the larger moons are composed mainly of ice with rocky interiors, but

Figure 1.1. Part of the surface of Ganymede, a satellite of Jupiter, showing a surface scarred by a variety of geological processes.

Table 1.1 The satellites of the outer planets, showing the characteristics of all near spherical satellites whose radius exceeds about 200 km. Pluto is included because of its resemblance to Triton. Data for Earth, the Moon, and the other terrestrial planets are included at the foot of the table for comparison. (For satellites with elliptical orbits, the orbital radius quoted is the semimajor axis.)

Planet	Satellite	Radius (km)	Mass (10^{20} kg)	Density (10^3 kg m^{-3})	Orbital radius (10^3 km)
Jupiter	Io	1815±5	894 ±2	3.57	421.6
	Europa	1569±10	480 ±2	2.97	670.9
	Ganymede	2631±10	1482.3 ±0.5	1.94	1070
	Callisto	2400±10	1076.6 ±0.5	1.86	1883
	12 others	<135			
Saturn	Mimas	197±3	0.38±0.01	1.17	185.52
	Enceladus	251±5	0.8 ±0.3	1.24	238.02
	Tethys	524±5	17.6 ±0.3	1.26	294.66
	Dione	559±5	10.5 ±0.3	1.44	377.40
	Rhea	764±4	24.9 ±1.5	1.33	527.04
	Titan	2575±2	1345.7 ±0.3	1.88	1221.85
	Iapetus	718±8	18.8 ±1.2	1.21	3561.3
	15 others	<175			
Uranus	Miranda	236±3	0.75±0.22	1.35±0.39	129.8
	Ariel	579±2	13.5 ±2.4	1.66±0.30	191.2
	Umbriel	586±5	12.7 ±2.4	1.51±0.28	266.0
	Titania	790±4	34.8 ±1.8	1.68±0.09	435.8
	Oberon	762±4	29.2 ±1.6	1.58±0.10	582.6
	12 others	<85			<86
Neptune	Triton	1350±5	213.8 ±1	2.075±0.019	354.8
	7 others	<200			
Pluto	—	1145–1200	131.5 ±0.3	1.92–2.06	—
	Charon	600–650	15.6 ±0.3	1.51–1.81	19.6
Mercury	—	2439	3300	5.4	—
Venus	—	6051	48700	5.3	—
Earth	—	6371	59970	5.517	—
	Moon	1738	734.9 ±0.7	3.34	384.4
Mars	—	3394	6420	3.9	—

two of them (Io and Europa) have silicate rock reaching outward to, or almost to, their surfaces. Only Io, Europa, and Ganymede appear to have an Earth-like iron-rich core. With the exception of Titan, atmospheres are tenuous or absent, so the surface is unprotected from cosmic radiation. For those satellites orbiting within their planet's magnetosphere, this effect may be made more severe by the enhanced concentration of charged particles. In general, the geological story of each of these worlds is written on a surface consisting essentially of ice. This does not mean that the outer solar system is a playground for glaciologists. The lucky people are geologists in general, because at the low temperatures prevailing so far from the Sun, the ice behaves mechanically very much like silicate rock, both at the surface and at depth. Furthermore, in the outer reaches of the solar system the ice is not usually pure water-ice (H_2O); it is contaminated by such things as ammonia (NH_3) and methane (CH_4), which cause it to melt in ways analogous to

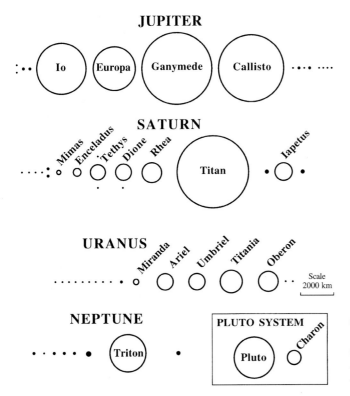

Figure 1.2. The satellites of the outer planets, showing their relative sizes and ranged from left to right in order of distance from each planet. Names are shown only for the near-spherical bodies, greater than about 200 km in radius, which are the main topic of this book.

the partial melting of terrestrial rocks and give rise to the possibility of a comparable range of volcanic and intrusive "igneous" processes. In this book, and in other literature on the subject, the term "ice" is used to refer to any frozen volatile substance or mixture.

Some of the highlights of the outer planet satellites are introduced in the next few pages. Chapter 2 then takes a look at the origin and evolution of these worlds and reviews what we knew about them before the *Voyager* missions. Chapter 3 gives an overview of the remarkable *Voyager* project and the subsequent *Galileo* orbital mission to Jupiter. In chapter 4 the ground rules are presented for understanding the geology of the larger moons in terms of their internal structure; they are layered bodies, and in particular the concept of a rigid icy outer layer (the lithosphere) overlying a weaker icy layer (the asthenosphere) is set up by analogy with Earth's lithosphere and asthenosphere, which are made of silicate rock. The surfaces of the moons themselves are discussed in detail in the succeeding chapters, and the final chapter looks ahead to future missions. I have kept the vocabulary as simple as I could, and the unavoidable technical terms are defined in the glossary.

1.2 A GALLERY OF WORLDS

Those moons of the outer planets that have not been extensively resurfaced by geological processes are characterized by vast numbers of impact craters. These

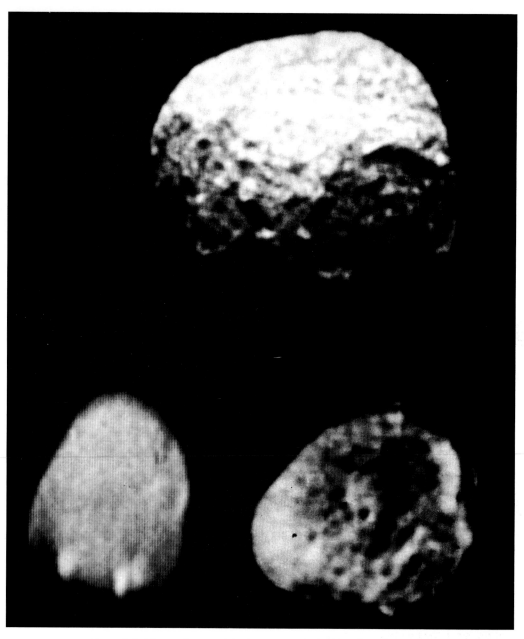

Figure 1.3. Views of some of the larger irregularly shaped satellites. These are perhaps fragments resulting from the collision of larger bodies. Moons such as this are too small for geological processes and are not considered in any detail in this book. (*Top*) Proteus, a satellite of Neptune discovered in 1989 by *Voyager-2* (semimajor axes 218 × 208 × 201 km), whose own gravity is not quite sufficient to have pulled it into a spherical shape. (*Lower left*) Amalthea, an inner satellite of Jupiter (semimajor axes 135 × 82 × 75 km). (*Lower right*) Hyperion, an outer satellite of Saturn (175 × 120 × 100 km).

Figure 1.4. Part of the surface of Rhea, a satellite of Saturn whose surface is saturated by impact craters. This view shows an area about 300 km across.

craters may occur in such profusion that they completely cover the surface, leaving no trace of its former state, as shown, for example, in Figure 1.4. Most of the highlands of Earth's own Moon are densely cratered in this way, but younger terrains have fewer craters, and variations in the numbers and sizes of craters from place to place are used to estimate the ages of the various units exposed on the surface.

Such dating using cratering statistics works well in the inner solar system, as far as can be told with the limited independent (radiometric) dates from the Moon and none from the other planets. The standard model is that the oldest surfaces seen, those virtually saturated with craters, date from about four billion years ago. This was the time of the so-called late heavy bombardment by fragments left over from the formation of the planets 500 million years before. The supply of impacting bodies began to decline rapidly after about 3.9 billion years ago and has remained at a broadly uniform rate (though perhaps interrupted by a few flurries) during the past 3.8 billion years. Thus, for surfaces younger than 3.8 billion years there is a fairly straightforward correlation between age and density of craters. As you will see in chapter 5, in the outer solar system the calibration between the age of a surface and its crater density is less certain, because there are alternative populations of impacting bodies available and heavy cratering may have occurred much more recently than it did in Earth's neighborhood. Nevertheless, the older a surface is, the more craters it will tend to have. A surface that is densely cratered cannot have been subject to widespread resurfacing by geological processes such as volcanism or tectonism (faulting and other deformation) since the rate of cratering declined.

To the geologist, then, the heavily cratered moons are the least interesting ones. However, even these have tales to tell. For example, the cross-sectional shapes of the craters, particularly the big ones, can tell us about the strength of the outer layers of the moon. If the topography of a crater is subdued and appears to have become flattened by subsidence, then it can be assumed that the outer layers are (or were once) thin or weak enough to deform under their own weight. This information can be used to constrain models of the moon's internal structure and thermal history. Chapter 5 shows that such properties are far from uniform even between moons that appear, at first sight, to look much the same.

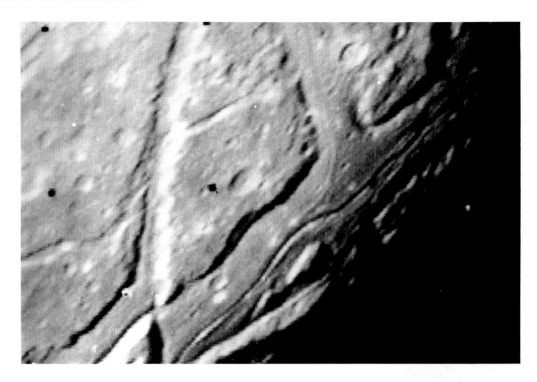

Figure 1.5. A 400 km wide view of part of Ariel, a satellite of Uranus, showing ancient, heavily cratered terrain cut by a major fault-bounded valley. The valley floor has been filled by a volcanic flow unit that is evidently much younger than the surrounding terrain because of the comparative sparseness of craters in this area.

On many of the other moons, terrains preserving evidence of a period of dense cratering have been wholly or partly covered over or replaced by younger surfaces formed either in response to faulting or as a consequence of heating or resurfacing by volcanic processes (Figs. 1.5 and 1.6). The height and steepness of fault scarps and other slopes can be used, just like craters, to give information on the strength of the outer layers of the moon. As discussed in chapters 4, 6, and 7, the volcanism that has occurred is generally the eruption of ice-dominated flows, but in ways closely analogous to many terrestrial volcanic processes.

Whether the dominant resurfacing agency has been tectonic or volcanic, moons that show signs of such geological activity have evidently had enough internal energy at some time to drive these processes. This begs the question of what was the heat source. The small amount of rock in these bodies makes radiogenic heat (from the decay of radioactive elements), which drives Earth's heat engine, an unlikely candidate except during the extreme youth of the solar system. Therefore, other sources have to be considered, such as energy derived from gravitational interactions through tidal processes. The amount of energy required to cause the observed amount of resurfacing depends on how the composition, strength, and temperature of the body vary with depth, and so the observations that can be made of their surfaces have a great deal to tell us about how each of these moons has evolved. We shall return to this theme in chapters 6 and 7.

The jewels in the crown of solar system exploration have been those worlds where geological processes have actually been observed in action. The most remarkable among these are Io and Triton. Io was one of the first moons to be imaged by *Voyager-1*. Blending geophysical prowess and impeccable timing in equal measure, Stanton Peale and colleagues proposed a model predicting that

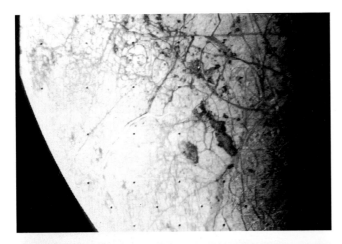

Figure 1.6. A 2000 km wide view of part of Europa, a satellite of Jupiter. The surface is cut by a variety of tectonic features, and because there are very few impact craters it is evidently very young.

Figure 1.7. The image on which the first volcanic plumes on Io, a satellite of Jupiter, were discovered. The bright spot to the left of the crescent disk is a plume rising into the sunlight from the Loki fissure, and the top of a fountainlike second plume is faintly visible rising some 300 km above the bright limb of Io, toward the lower right, from the Pele eruptive center.

the interior of Io should be largely molten because of tidally generated heat, which was published in the journal *Science* just a few days before *Voyager-1*'s encounter with Io in 1979. Confirmation of Io's internal heat was dramatically provided by *Voyager* images showing not only a surface dominated by fresh-looking volcanic features that obliterate all traces of impact craters, but also nine active volcanoes with eruption plumes reaching over 300 km in height (fig. 1.7). Remapping of Io by *Galileo* beginning in 1996 showed that several of

Figure 1.8. General view of Triton, a satellite of Neptune, showing a variety of terrains. The south polar cap, overlain by dark streaks representing dust deposits from wind-blown eruption plumes, occupies the left of the image. A color view of the same region is shown in plate 7.

these volcanoes were still active and surrounded by fresh deposits, and that other volcanoes that had been inactive during the *Voyager* encounters had since come to life.

Triton was the last object in the solar system to be visited by *Voyager-2*. Some people had hoped to find oceans of liquid nitrogen. They were frustrated in that respect, but the surface contains evidence of such a wide variety of tectonic, volcanic, and other geological processes that Triton was far from a disappointment. Triton has arguably the most fascinating, and least-well understood, surface of all the outer planet satellites (fig. 1.8). In addition to this, the images showed several geyserlike plumes that reached an altitude of 8 km before being driven downwind in the tenuous atmosphere.

Io, Triton, and two other probably active moons (including Europa, revealed in superb detail by *Galileo*) are discussed at length in chapter 7. Chapter 8 presents what we know about the geology of Titan, the second largest satellite in the solar system (which is not very much because Titan's atmosphere hid its surface from *Voyager*'s gaze). It also discusses the planet Pluto (which may be similar to Triton), its satellite Charon, and the myriad of icy bodies in the trans-neptunian Kuiper belt whose existence remained unproven until the 1990s. For now, however, you are invited to come down to Earth and consider the pre-*Voyager* view of planetary satellites and their significance in understanding the formation and evolution of the solar system.

2   Composition and Evolution of Satellites

And now a bubble burst, and now a world
Pope,
An Essay on Man

When Jupiter is at its closest to Earth (just over 600 million km), the largest of its satellites appears less than two seconds of arc across. This is about as big as your thumbnail seen from a distance of three kilometers. Satellites of the other outer planets are even farther away and physically smaller, so it is not surprising that ground-based observations have been able to provide so little information about their geology.

Thus, in a 550-page astronomical text written near the end of the nineteenth century, the author has this to say about the surfaces of the galilean satellites:

> Owing to the minuteness of the satellites as seen from the earth, it is extremely difficult to perceive any markings on their surfaces, but the few observations made seem to indicate that the satellites (like our moon) always turn the same face towards their primary. Professor Barnard has, with the great Lick refractor, seen a white equatorial belt on the first satellite, while its poles were very dark. Mr. Douglass, observing with Mr Lowell's great refractor, has also reported certain streaky markings on the third satellite. (Sir Robert S. Ball, *The Story of the Heavens*, Cassell & Co., first published 1886)

The Lick refractor had a 90 cm aperture, and Lowell's instrument (with which he made his famous but illusory observations of the canals of Mars) had an aperture of 61 cm; these are among the biggest and optically most precise conventional refracting telescopes ever made, and the nature of the observations that were made with them indicates the limitations of ground-based observations in mapping satellite surfaces. Not surprisingly, the same author has nothing at all to say about the surfaces of Saturn's satellites.

15

So far as they go, the descriptions just quoted of the pale equatorial region on Io (the first satellite) and the general streaky nature of Ganymede (the third satellite) are now known to be broadly correct, but they did not cast much light on what the surfaces of these worlds are really like. However, the realization that these bodies are in synchronous rotation, so that they rotate on their spin-axis once during the course of each orbit, was an important discovery. In the next section we discuss what this and other observations that can be made from the ground tell us about the nature of planetary satellites.

2.1 PLANETARY SATELLITES
SEEN FROM EARTH

We now know that all the outer planet satellites greater than about 200 km in radius are in synchronous rotation. This can hardly be coincidence. What it implies is that the gravitational, or tidal, influence of the planet around which each one orbits has slowed the satellite's original rotation down until it matches its orbital period. The process is understood to have operated as follows. The gravitational attraction of the planet distorts the shape of the satellite by raising tidal bulges on it. Ideally, one of these bulges should always be centered on the hemisphere facing the planet, with the other bulge exactly opposite. In a real material there is no way in which the bulges can keep pace if the satellite is spinning rapidly. They therefore lag behind the direct alignment, just as high tides in Earth's oceans are out of phase with the position of the Moon in the sky. The tidal force exerted by the planet on the bulges eventually drags them into line, which it can do only by slowing down the satellite's rotation until it matches the period of the orbit. Once this has been achieved the bulges have become stationary on the surface of the satellite, and, if the satellite were in a circular orbit, no further work would be required to keep them in place.

All this is presumed to have happened early in each satellite's history. The calculated timescales for bringing most satellites into synchronous rotation are in the range of ten thousand to ten million years. A crucial consequence of this tidal "despinning" is that it must have generated a large amount of heat within the satellite. As we will see, the magnitude and duration of such heating events have important consequences for the development of internal layering and the operation of tectonic and volcanic processes at the surface.

Telescopic observations also enabled the masses of several of the satellites to be measured from calculations based on the perturbations they cause to one another's orbits—so the extremely detailed orbital data referred to in chapter 1 came in handy after all! Masses that were derived for Io, Europa, and Ganymede by this method in 1921 are within five percent of their currently accepted values. However, to work out what the satellites are likely to be made of, it is important to know their densities. The only ground-based way to determine density is to calculate it from mass and volume. The angular size of a planetary satellite seen from Earth is so small that it is difficult to measure precisely, and because volume depends on the cube of the radius, the errors in volume, and hence in density, determined by ground-based observations were several times greater than those for mass. Nevertheless, it was possible to demonstrate that the inner galilean satel-

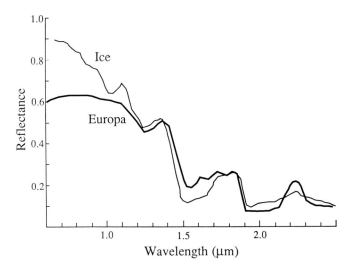

Figure 2.1. The reflectance spectrum of Europa compared with that of coarse-grained water-ice, measured between 0.6 and 2.5 µm wavelength. The visible spectrum extends only as far as 0.7 µm, so this figure shows mostly the reflected infrared part of the spectrum. The dips in the spectra at 1.04, 1.25, and 1.5 µm are absorption bands that could be caused by H_2O or -OH, but the absorption band at 2.0 µm is characteristic of H_2O alone.

lites (Io and Europa) are denser than the outer two (Ganymede and Callisto). This is a striking parallel with the solar system itself, where the inner planets are denser than the outer ones, a theme to which we shall return later.

Even knowing the density of a satellite does not tell us what it is made of. A second important line of evidence that can be gathered from ground-based observations—and this is the one that enables us to rule out the proverbial green cheese and other fantastic materials—is the spectroscopic analysis of the light reflected by these bodies. When matter reflects light, the molecular bonds within it absorb certain wavelengths, resulting in a series of absorption bands in the reflected spectrum that are characteristic of the composition of the surface. Thus, by measuring the way that the intensity of sunlight reflected by a satellite varies with wavelength, it is possible to understand the composition of the surface material. To make sure that only the absorptions caused by substances on the satellite itself are measured, it is necessary to make a correction for the shape of the Sun's spectrum and also to allow for absorption by gases in Earth's atmosphere above the observatory.

The first breakthrough with this technique came in 1944, when Gerard Kuiper found absorption features due to methane in the reflectance spectrum of Titan. As Kuiper realized, the particular features he saw could be caused only by methane in the gaseous state, which proved that Titan (alone among the planetary satellites, as is now known) has a substantial atmosphere. In fact, the atmosphere of Titan is so opaque that it prevents the surface being seen using visible light, and its nature is still unclear (chapter 8). There is no such problem with the other satellites, and from the late 1950s onward Kuiper and his successors were able to demonstrate the presence of water-ice dominating the visible and near infrared reflectance spectra of many of the other major satellites (fig. 2.1). A notable exception is Io, where there was evidently little or no water; instead, Io has a highly reflective, markedly red surface, which during the 1970s some researchers began to attribute to the presence of sulfur.

The detection of water-ice on most of the planetary satellites was certainly compatible with the very low surface temperatures (120–170 K) of the galilean satellites, first revealed during the 1960s by the use of newly developed thermal infrared telescope instruments. However, the trouble with relying on spectral information is that the compound that dominates the spectrum is not necessarily the most abundant one on the surface; in addition, the technique only gives information about, at most, the outer few millimeters of the satellite. What if it has a thin coating that is not typical of the body as a whole? Thermal infrared determination of the rate at which satellites cool as they pass into the shadow of their planet shows that they generally have a low thermal conductivity, so at least the outermost meter or so is likely to consist of loose fractured and broken material. An unconsolidated powdery nature to their surfaces can also be demonstrated by measuring how they scatter sunlight in different directions. Both observations are consistent with surface material physically resembling the lunar regolith of impact-generated debris in which the *Apollo* astronauts left their footprints. If the regolith on the satellites is the result of an impact process (as the cratering we have since observed on most of them indicates), then this suggests that there must be a considerable degree of mixing and hence homogenization of the outermost material to some depth. Unfortunately, this does not rule out the presence of frosts and other surface coatings that could dominate the spectrum.

Despite the reservations quoted above, by the early 1970s it had become fairly clear that, with the exception of Io and Europa, the satellites of the outer planets for which there were reasonable data have fairly low densities, in the range of 1.2–2.0×10^3 kg m^{-3}. This is less than twice the density of water, and only about half the density of the rock that makes up Earth's crust. A reasonable assumption at the time, which is still regarded as good, is that these bodies are composed of a mixture of rock and ice. The domination of their reflectance spectra by water-ice combined with the fact that their density is generally greater than that of ice is a good indicator that their exteriors are icy and their interiors rocky.

This concept was (and remains) compatible with models of how the satellites were assembled, and the nature of the material from which they could have formed early in the life of the solar system. It is worth pausing briefly, before following the story of the exploration of the planetary satellites any further, to review what we can infer about their origin and early evolution.

2.2 ORIGIN OF PLANETARY SATELLITES

This section gives an overview of the origin and evolution of planetary satellites. It is a post-*Voyager* view, although in many of its essentials it resembles models that were developed during the 1970s. To understand how the planetary satellites were formed, we must first consider how the solar system as a whole developed.

2.2.1 Origin of the Solar System

It is almost universally accepted that the Sun and its planets grew from a cloud of gas and dust. As this cloud contracted under its own gravity, much of its mass became concentrated at the center, in the proto-Sun. The original cloud was rotating, and as its center contracted, the conservation of angular momentum

Table 2.1 Solar (cosmic) abundances for the ten most common elements, normalized (by convention) to Si = 1.0×10^6. These abundances are assumed to have been much the same in the solar nebula.

Element	Abundance
H	28.2×10^9
He	1.8×10^9
O	16.6×10^6
C	10.0×10^6
N	2.4×10^6
Ne	2.1×10^6
Si	1.0×10^6
Mg	850×10^3
Fe	700×10^3
S	460×10^3

forced the rest of the cloud into a disk, which was flattened and rotating in the same plane as the Sun itself. This disk is usually called the solar nebula. It probably took not much more than ten thousand years for this process to occur. By the end of that time nucleosynthesis had started within the Sun and it was shining hotly. Various very young stars in this stage of evolution, surrounded by disks of gas and dust, have been recognized particularly with the aid of infrared astronomy.

The inner parts of the solar nebula were hot, even before the Sun's nuclear reactions began, because the proto-Sun would have been radiating strongly as a result of gravitational energy lost during contraction. With the passage of time, the solar nebula became less dense as a consequence of two processes: the loss of material to interstellar space and condensation to form solid objects. This lower-density nebula allowed heat to be radiated away more easily, leading to a general cooling of what remained of the nebula. At some time during this cooling and dissipation stage, the material that now forms the planets condensed and aggregated from the nebular material.

So what was this solar nebula made of? Fortunately we can determine its initial composition quite easily, if we are prepared to accept that the outer layer of the Sun has remained an unadulterated sample of this material. As the Sun is so hot, its elemental abundances can be measured spectroscopically simply by measuring the depth and wavelength of the absorption bands in its spectrum. The solar abundances of the most common elements are listed in table 2.1. Primitive meteorites known as carbonaceous chondrites have similar elemental abundances (excluding gases), which greatly strengthens the argument that both they and the Sun are reliable samples of solar nebula material.

Condensation of much of the solar nebula material probably took place over an interval of about ten thousand to a million years. As indicated in table 2.2, the earliest material to condense was composed of the most refractory molecules, that is, those having the highest condensation temperatures. These include elements that are not in the top ten of the abundance charts, such as calcium, aluminium, and titanium occurring as oxides and silicates. We have direct evidence of this in the form of tiny grains of these substances that are found as inclusions

Table 2.2 The temperatures at which some notable constituents of the solar nebula may have begun to condense. Hydrogen, neon, and helium could have condensed only below 30 K, and their presence in the gas giant planets is almost certainly due to gravitational capture as gases; these elements are rare in smaller bodies, including planetary satellites.

Temperature (K)	Mineral
1758	Corundum, Al_2O_3
1647	Perovskite, $CaTiO_3$
1513	Spinel, $MgAl_2O_4$
1471	Nickel–iron metal, Fe, Ni
1450	Diopside (pyroxene), $CaMgSi_2O_6$
1444	Forsterite (olivine), Mg_2SiO_4
700	Troilite, FeS
550–330	Hydrated minerals
180	Water, H_2O
120	Ammonia, as $NH_3 \cdot H_2O$
70	Methane, as $CH_4 \cdot 6H_2O$
70	Nitrogen, as $N_2 \cdot 6H_2O$

in many meteorites. As the temperature of the nebula decreased, nickel and iron condensed, followed by more familiar rocky minerals, such as magnesium-rich silicates like olivine and pyroxene. Much of this matter may have collected into globules from about a millimeter to a centimeter across, now preserved as "chondrules" embedded in primitive meteorites (fig. 2.2). Radiometric dating shows that chondrules formed over an interval of no more than a few hundred thousand years at about 4.5 billion years ago, which is normally regarded as the age of the solar system. At lower temperatures, progressively less refractory (i.e., more volatile) constituents of the solar nebula condensed, such as iron sulfides. Of the more abundant elements in table 2.1, only oxygen had been a significant contributor to the condensates up to this point, but at lower temperatures it became possible to begin to mop up the hydrogen (in water, H_2O), nitrogen (in ammonia, NH_3), and carbon. Perhaps the greatest remaining uncertainty from the point of view of interpreting the composition of bodies in the outer solar system is the form in which most of the carbon resides. This depends on the temperature at which the atoms in the solar nebula equilibrated to form molecules, and on the degree of availability of oxygen at the time, with the result that most of the carbon could have been taken up either in methane (CH_4) or carbon monoxide (CO). To complicate the issue further, some models call for a proportion of elemental carbon (e.g., graphite) as a condensate, and it is clear from carbonaceous chondrite meteorites and studies of the nucleus of Halley's comet that some of the carbon ends up in organic compounds.

While this sequence of cooling and condensation was taking place, the condensed material was aggregating into progressively larger masses. At first this was because those particles whose random motion brought them into contact with each other tended to adhere, but eventually the gravitational attraction of the larger of the planetesimals that formed in this way enabled them to vastly outpace the rate of growth of their fellows.

Figure 2.2. Close-up of a fragment of the Allende meteorite, which is a carbonaceous chondrite. Roughly spherical chondrules, which may represent originally liquid droplets, are embedded in a carbon-rich silicate matrix. The large pale chondrule above and left of the center is about 2.8 mm across.

Probably the most important unresolved question about the whole process of planetary accretion is the rate at which it occurred relative to the rate of cooling of the solar nebula. If the rate of cooling was slow, then the planets could have accreted heterogeneously, because the more refractory material would have had time to condense into grains, aggregate into planetesimals, and accrete into protoplanets before the more volatile material became available. On the other hand, if the rate of cooling was relatively fast, then there could have been dust particles with a whole range of compositions coexisting at one time, so the planets would have accreted in a homogeneous fashion. On the heterogeneous accretion model, the observed layering of the planets (dense, mainly refractory, interiors with less refractory outer layers, surrounded by a hydrosphere and atmosphere) is a natural consequence of sequential accretion. On the homogeneous accretion model, the layering is a result of later gravity-driven differentiation and the outward migration of volatile material. Either scenario can be adapted to explain the generally decreasing densities of the inner planets (Mercury, Venus, Earth, and Mars) outward from the Sun and their deficiency in light, gaseous elements compared with the outer giant planets, by stipulating that temperatures within the solar nebula in the inner solar system remained too high for the more volatile constituents to condense. For example, it is unlikely that water-ice could have condensed directly from the solar nebula any closer to the Sun than about five times Earth's orbital radius (i.e., 5 astronomical units, or AU).

These homogeneous and heterogeneous accretion models represent extreme variants of what could have taken place. It is generally accepted that the truth lies somewhere between the two. What then does this tell us about the satellites of the outer planets? Treated at its simplest, it suggests that, if they formed roughly where they are now seen, they should contain progressively more volatile material as the distance from the Sun increases. As you will see in chapters 5–7, this is compatible with the evidence that the water-ice is contaminated with volatiles such as ammonia by the time the Uranus system is reached, with the addition of methane and nitrogen in the Neptune system. The first appearance of ice-dominated bodies at Jupiter is compatible with the 5 AU inner limit for the condensation of water, as Jupiter's orbit has a radius of 5.2 AU.

2.2.2 Satellite Formation

But how did the satellites form? Was their creation linked to the formation of their planet, or did they form independently and become captured later? One of the most important pieces of evidence in this respect is that all the major planetary satellites except Triton orbit their planet in the same direction as the planet's spin. The inclination of their orbit to the plane of their planet's equator is very small, usually less than 1°. Capture of all these moons into orbits of this nature is unlikely, and for Jupiter, Saturn, and (perhaps) Uranus, the major satellites are generally regarded as having been formed by condensation from a disk of gas and dust around each primitive planet, each rather like a smaller version of the original solar nebula. Such a disk could have been formed by material that was shed by the protoplanet as it shrank. More likely, it grew around it. This could occur by scavenging gas and dust from the solar nebula, by collecting debris from collisions between planetesimals, or as a result of collisions of large planetesimals with the protoplanet that sprayed debris into orbit. Either way, the temperature would be likely to have increased toward the center of the disk, providing an opportunity for the chemical composition of the satellite-forming material to evolve.

Whatever the origin of such a protosatellite disk about a planet, it must have developed into a satellite system in much the same way. Small particles would have aggregated into larger ones that over time would have collided. Eventually the largest bodies would have swept up all the debris in their vicinity to form the moons we see today. The amount of heat, and consequent melting, liberated in this accretion process must have been an important factor in controlling the thermal evolution and differentiation of the satellite, and this is an issue to which we return in chapter 4. The outward decreasing density of the galilean satellites of Jupiter is usually taken to imply that there was less capture or retention of volatiles closer to the proto-Jupiter, as a result of elevated temperatures in this region caused by thermal radiation from the still-contracting planet. As remarked earlier, in this respect the Jupiter system mimics the solar system in miniature. However, no such progression in densities is evident among the moons of Saturn or Uranus (table 1.1), and some or all of these satellites may have long and complex histories, so they no longer retain their original size or mass distribution.

There are various exceptions to this model of the origin of satellites. The first of these are most of the small satellites of the outer planets. Some of these are icy bodies with near-circular orbits, which may represent fragments from collisions involving larger moons. Their rocky or carbonaceous fellows, and those that are icy but are in strange orbits, may equally well be captured asteroids or comet nuclei. The second prominent exception is Triton, the only really large moon of Neptune. This has a retrograde orbit, meaning that its orbital motion is in the opposite direction to the planetary spin. Moreover, the orbit is inclined to the planet's equator at a considerable angle (21°). The most probable explanation is that Triton formed elsewhere but was captured by Neptune's gravity after a close encounter with the planet. This idea is supported by the fact that, as far as we can tell, the planet Pluto is more like Triton than anything else in the solar system. Moreover, there are now known to be dozens (and suspected to be thousands or even millions) of icy bodies ranging from a few hundred kilometers in size downward orbiting the Sun close to or beyond the orbit of Neptune, mostly at about 30–50 AU from the Sun. Their existence was predicted by Kenneth Edgeworth in the 1940s, and again by Gerard Kuiper in 1951, on the grounds that there ought to be planetesimals that had been prevented from growing into larger bodies because of the infrequency of collisions in the vast reaches of trans-neptunian space. The region became known as the Kuiper belt (sometimes the Edgeworth-Kuiper belt), long before its first inhabitant (other than Pluto and its satellite) was discovered in 1992. Pluto and Triton are probably the largest bodies to have grown in the Kuiper belt. There is no satisfactory explanation as to how Triton's capture may have occurred, but as you will see in chapter 7, this event may also have been the cause of some of the strange geological events that have taken place on it.

The Uranus system poses similar problems. The planet has five major satellites (table 1.1), all of which are in near-circular prograde orbits in the plane of the planet's equator. So far, so good: it looks not unlike Saturn's family of moons. However, the whole system is tilted on end, so that the axis of the planet's rotation lies at an angle of 98° to the plane of its orbit (another way of looking at this is to say that it rotates in a retrograde sense, on an axis inclined at 82°). For comparison, the axes of the three other giant planets are inclined at less than 30° to their orbits, and Earth's axis is tilted at a little over 23°.

It is likely that the rotation axis of Uranus originally had a similar orientation to that of the other planets, because it should have inherited its rotation from the nebula out of which it condensed. One way for Uranus to have been knocked over at some stage after its formation is for a large body (bigger, e.g., than any of its present satellites) to have collided with it. Such an impact could have created satellite-forming debris considerably later in the planet's history than most other models require. Interestingly, there are geochemical grounds for supposing that such a giant impact was responsible for the origin of the Moon, by condensation of the material knocked out of the mantle of the primitive Earth, but the grounds for demonstrating a similar process at Uranus are much weaker. However, it is worth noting that if the moons of Uranus existed before the planet's tilt was changed, this reorientation, however it happened, would almost certainly have badly disrupted them. Tidal processes can be called upon to drag the satellites' orbits back into line with the new rotation axis, but only over a prolonged period during which it is likely that there would have been several collisions between satellites or between satellites and large fragments from disrupted satellites. The resulting debris would have led to a new interval of satellite accretion.

The matter of what happens to an icy moon after it accretes, and in particular how it heats up internally and where and when melting may occur, was the subject of some thought-provoking studies during the final pre-*Voyager* years. These, and some developments arising from them, are discussed in the next section.

2.3 HEATING AND DIFFERENTIATION OF AN ICY MOON

To understand this topic requires a simple appreciation of planetary heat sources and the properties of ice. Let us assume for now that a planetary satellite accretes homogeneously; that is, it grows by the collision of lumps composed of an ice–rock mixture. The densities of most of the satellites (Io and Europa excepted) indicate that the mixture was fairly close to the cosmic abundance ratio of ice to rock, which is about 60:40 by mass. In this chapter we will usually keep things simple by regarding the ice as being just water-ice, but you should be aware that adding ammonia or methane to the ice will slightly decrease its density and, in the case of ammonia in particular, have important consequences in lowering the melting temperature and also making it easier for the ice to convect in the solid state. These matters are discussed further in chapter 4.

An important consideration is that when the volatile compounds listed in table 2.2 condense in a near-vacuum, they do so at temperatures considerably below their normal melting point, and go directly from vapor to solid. This means that in a homogeneously accreted satellite, in the absence of a heat source, there would be no melting and hence no way in which the rock could separate from the ice. The body would therefore retain its initial mixed, or undifferentiated, structure.

2.3.1 Accretional Heating

Perhaps the most obvious way in which a satellite forming in this manner can warm up is by the kinetic energy liberated by the impact of the accreting bodies. How much of this energy is retained within the satellite and how much is immediately radiated away to space is debatable; probably only about 10–50 percent is retained. Many models put forward during the 1970s and 1980s tended to agree that, because they are so massive, Ganymede and Callisto should have been warmed sufficiently by accretional heating to have melted by the time their radius had grown to more than about 1000–2000 km. This would have freed the rock particles and fragments embedded within the ice, which would then have sunk and accumulated at the center of the body. The result would be a differentiated, layered, structure with a rocky interior surrounded by a rock-free ice layer. Incidentally, the sinking of the heavier fragments would itself contribute about a further 10 percent to the heating, through the loss of gravitational potential energy. Once the input of accretional heat had ceased, the water would freeze from the outside inward, as heat was radiated to space. Note that the basic layered structure arrived at in this way is the same as that produced by heterogeneous accretion, in which the rock would have accreted first and the ice later.

Similar models for smaller bodies such as the moons of Saturn (Titan excepted) show fairly convincingly that these would never have collected or retained enough accretional heat for melting to occur. To predict the thermal evolution of these worlds, and to model the postdifferentiation history of larger bodies, other heat sources need to be considered. Earth is hot inside primarily as a result of the decay of radioactive elements, and this mechanism is another way in which the planetary satellites could have been heated.

2.3.2 Radiogenic Heating

Evidence from some meteorites suggests that they were heated very soon after they formed, by the energy released from the decay of ^{26}Al, a short-lived isotope of aluminium with a half-life of just 7×10^5 years. The only known way for ^{26}Al to form is by nucleosynthesis in a star, and it is presumed to have been dispersed into space by a supernova explosion or similar catastrophe shortly before the formation of the solar system. As a result of its short half-life, all but the minutest trace of ^{26}Al has long since disappeared from the solar system, although decay products from ^{26}Al are found in many meteorites in amounts that demonstrate that these meteorites contained ^{26}Al when they were created. This means that the solar nebula must have formed and condensed within just a few half-lives of the event that distributed the ^{26}Al. However, most researchers suppose that the formation of the planetary satellites, from meteoritic and other debris, occurred too long after the event for enough ^{26}Al to have survived for it to have played a significant role in even their initial heating.

The other heat-producing radioactive elements in solar system rocks have much longer half-lives. The major source of postaccretionary heat in Earth, the Moon, and the other terrestrial planets is from the decay of isotopes of uranium (^{235}U and ^{238}U), thorium (^{232}Th), and potassium (^{40}K). The half-lives of these isotopes are of the same order of magnitude as the present age of the solar system and they survive in appreciable quantities. The rate of radiogenic heat production from these must have been significantly greater when the solar system was young, and in rock of chondritic composition heat from ^{40}K should always have exceeded the heat from the other three heat-producing isotopes put together. It is natural to expect these elements to occur only in the rocky fraction of a planetary satellite, but potassium in particular could also occur in salt crystals dispersed within the ice as a result of chemical reactions between water and rock.

Models for the rate of radiogenic heat production in the rocky fraction of icy satellites usually assume that it has concentrations of these elements similar to the concentrations we can measure on Earth, on the Moon, and in chondritic meteorites. Taking account of the rate at which such heat would have been dissipated, the models generally show that moons less than about 600 km in radius are unlikely ever to have become warm enough for internal melting to have occurred. This means that if these satellites accreted homogeneously, they must have remained undifferentiated unless an additional heat source operated. The same models suggest that it would have been possible for the ice in the interior of a larger moon to melt, but because of the decline in radiogenic heat production over time, the layer of liquid water would subsequently have begun to freeze at its upper boundary.

The consequences of one such model, dating from 1978, for a moon of 700 km radius are shown in figure 2.3. For a body this small, the accretional heating is negligible, so at the time of formation the temperature is shown as uniform throughout (t_0 in fig. 2.3). Radiogenic heat from within the satellite is unable to escape as fast as it is generated, so the interior warms up. It begins to melt after about 0.6 billion years and reaches a maximum temperature at 2.0 billion years (t_1 in fig. 2.3). At this stage a core of rock has formed, by the sinking of the rock within the melt. The layer above is liquid water, which is kept at a uniform temperature (just above the freezing point) as a result of convection. Above the water is a layer that has not melted, and which therefore retains its initial composition. From this time onward there is cooling throughout as a result of declining radiogenic heat production, which is now exceeded by the heat loss to space. This re-

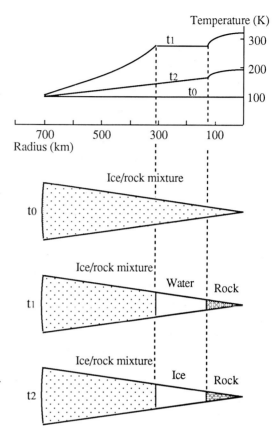

Figure 2.3. (*Top*) Thermal profiles through a radiogenically heated 700 km radius moon (60 percent water-ice, 40 percent rock). t_0, immediately after formation; t_1, after 2.0 billion years; t_2, after 4.5 billion years. (*Bottom*) Internal composition of the moon at the same times.

sults in the present temperature profile (t_2 in fig. 2.3), and in this example no liquid water layer remains. The internal structure has evolved through time as follows: initially an undifferentiated mixture of rock and ice (t_0); at the time of maximum melting (t_1), a 140 km radius core of rock, mantled by a 180 km layer of liquid water, overlain by a 380 km layer of undifferentiated rock and ice; at the present day (t_2), the same layered structure as the previous stage, except that the water layer has frozen right through.

This sort of model makes it reasonable to envisage the melting of a medium-sized satellite at some time in the past, which would have led to an at least partially differentiated structure, even if the accretion had been homogeneous. In larger moons, the melting would have reached closer to the surface before the radiogenic heating was outpaced by heat loss to space. However, any liquid water remaining would by now lie below a very great thickness of ice. It is doubtful whether radiogenic heating alone, even in the very largest moons, could have allowed differentiation within the rocky interior to yield a rocky outer core surrounding a denser iron-rich inner core.

We have already remarked on a third heat source for some of the planetary satellites earlier in this chapter, tidal heating, and we must return to this now to complete our survey.

2.3.3 Tidal Heating

As described in section 2.1, until the rotation of a satellite becomes synchronous with its orbital period, heat is generated internally. This is a result of the continual flexing of the moon as a whole that is necessary to allow tidal bulges to propagate around the globe. This could have been an important source of additional heat early in the life of a satellite, although its magnitude is not well constrained, or in the interval following capture in the atypical case of Triton. More generally, however, there is a significant source of tidally generated heat that continues to operate even today for some synchronously rotating satellites.

The rotation of a satellite on its axis is controlled by the conservation of angular momentum, and so the rate of rotation is effectively constant. However, the speed at which a satellite orbits its planet is described by Kepler's laws of planetary motion, which state that the speed is greatest when the satellite is closest to its planet. For a perfectly circular orbit, the orbital speed is uniform and will keep exactly in pace with the rotation of a synchronously rotating satellite. This means that the tidal bulges will always be aligned with the planet: no work needs to be done to try to keep them in line, so no tidal heat is generated. On the other hand, for a synchronously rotating satellite in an elliptical orbit, the changes in orbital speed mean that when the satellite is closest to the planet (and so moving the fastest), its rotation lags slightly behind exact synchroneity, and when the satellite is farthest away its rotation gets slightly ahead.

The result of this is a slight wobble, or libration, in the satellite's orientation as seen from the planet. Work is then expended to try to move the tidal bulges into step with this changing orientation. This causes internal heating. In the absence of outside influences, the eventual outcome would be gradually to nudge the satellite's orbit toward an exactly circular shape. However, this is not found in reality. The orbits of all the major satellites of the outer planets have measurable eccentricities, although these are less than about 0.02 in all cases (which means that when drawn on paper they look indistinguishable from circles). The reason for the persistence of orbital eccentricity is that the gravitational attraction between one satellite and another affects the shape of its orbit. The effect is particularly pronounced for the inner three galilean satellites of Jupiter because of the phenomenon of orbital resonance. For every one orbit made by Ganymede, Europa completes two orbits and Io four; this periodic alignment is enough to force their orbital eccentricities to remain far higher than they would otherwise be. The result is that their tidal bulges are continually flexed due to their changing orientation with respect to Jupiter's gravitational field. The amount of heat actually generated depends on the thickness and elasticity of layers within each satellite, for which estimates vary. In the case of Io, heat is generated by tidal dissipation at probably about 40–100 times the rate of radiogenic heating. This could be enough to melt at least some of Io's interior, and as noted in chapter 1, a prediction based on such a model proved spectacularly successful when *Voyager-1* discovered Io's active volcanoes. A similar kind of extra heating of certain icy satellites may have been important in the past, especially as such bodies require less heat input than rocky bodies before they melt.

2.3.4 Implications of Heating
and Differentiation

Thus there are three main mechanisms by which the interior of a satellite may be heated: accretional, radiogenic, and tidal. In reality, these will all have contributed to the thermal history of the body, although perhaps not sufficiently to have had any noticeable effect. In general, it is expected that the biggest satellites must have experienced more accretional and radiogenic heating than the smaller ones. The strength of tidal heating depends on proximity to the planet and the history of orbital resonances with the adjacent satellites.

One of the most important implications of the models discussed is that melting and the consequent differentiation may have affected only the interior of a moon, leaving an outer layer with the original primordial rock and ice mixture. Rapid cooling by radiation to space may even make it possible to retain an unmelted outer layer on bodies big enough to have melted throughout by accretional heating. In general, it is unlikely that the smaller major satellites ever experienced substantial melting, so if they accreted homogeneously they may still be largely undifferentiated.

There are various complicating factors and unknown parameters that make generalizations dangerous to use as guides for understanding the evolution of a planetary satellite. The factor that is perhaps the simplest to appreciate is that a mixture of rock and ice is denser than water, and so the outer layer of a partly differentiated body such as that in figure 2.3 at t_1 is gravitationally unstable. If the liquid water shell reaches close enough to the surface, the whole outer layer could fracture and founder, displacing the water toward the surface and leading to a reversal of the outer two layers. The result would be pure ice on the outside and mixed rock and ice in the middle layer.

Another uncertainty in the modeling is in the properties of ice at the high pressures and low temperatures that prevail within a planetary satellite. In particular, how ice deforms at very low (but geologically reasonable) strain rates cannot be measured under laboratory conditions. Seismic data show that Earth's mantle is solid, yet it is evident that convective forces cause it to flow at rates of the order of 1 cm yr^{-1}. In the simple model of radiogenic melting presented in figure 2.3 it was assumed that all the heat loss from the ice and ice–rock mixtures is by conduction. However, if the ice is able to flow at a rate comparable to the rate of flow of Earth's mantle, then this can transport heat outward more efficiently than by conduction. By removing internal heat in this way, solid-state convection by ice could inhibit or even prevent melting.

One of the major complicating factors is that the molecular arrangement of the crystalline structure of ice is dependent on temperature and pressure. The ice cube with which you may be familiar in your favorite drink is of the form known as ice I. If you were gradually to compress it (keeping it cold at the same time), there would come a point when its structure switched to an alternative, more closely packed arrangement of molecules. In fact, there are many different solid phases of water-ice, the stability fields of which are shown in figure 2.4.

Figure 2.4 is what is known as a phase diagram. The lines on it are the boundaries of the pressure–temperature conditions at which the various phases, or polymorphs, of ice are stable. Thus, at a temperature of 200 K, ice I will change

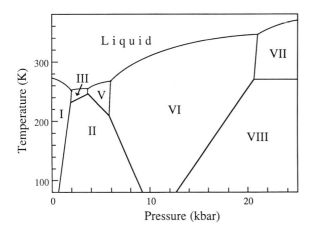

Figure 2.4. Simplified phase diagram for pure water-ice.

to ice II when the pressure is increased to about 1.6 kbar. The upper, curved line on the diagram, known as the liquidus, shows the conditions under which the ice will melt. Note that a sample of ice II cannot be melted directly. On warming it changes to another form of ice (ice I, III, V, or VI, depending on the pressure) before it reaches the liquidus.

The depth represented by a given pressure is different for every moon, because their different masses control their gravity. The depth–pressure relationship also depends on the average density, which is controlled by the exact proportions of rock and ice present, and the density gradient, which is affected by the degree of differentiation. Generally speaking, however, the pressure inside an icy moon smaller than about 600 km in radius will be low enough to allow ice I to be stable all the way to its center. In moons the size of Iapetus and Rhea (700–800 km in diameter), the pressure near the center is high enough to make ice II the stable phase, whereas the pressure–temperature gradient in a large moon such as Callisto is likely to take it in turn through the stability fields of ice I, ice II, ice VI, and then ice VIII, moving inward. Figure 2.5 shows two alternative models for Callisto, indicating the zones of stability of its ice phases.

The fact that ice has several phases that are stable under different conditions influences the evolution of an icy moon in two ways. In the first place, it is difficult for solid-state convection to occur across phase boundaries, so convective heat loss in the larger moons that have several internal phase transitions will tend

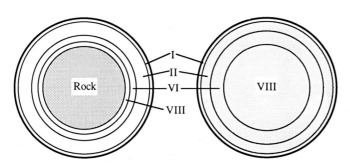

Figure 2.5. Differentiated (*left*) and undifferentiated (*right*) models for Callisto. The former has a silicate core (dense stipple) surrounded by pure ice; the latter has a mixture of rock and ice throughout. The ice phases that are stable in each zone are indicated by roman numerals.

to be inhibited. The second involves consideration of what happens to the solid part of a moon as it cools. In particular, a phase change from ice VI to ice VIII would occur within the larger satellites simply as a result of falling temperature. For example, at a pressure of 18 kbar ice VI inverts to ice VIII as soon as the temperature drops below 200 K. This phase change is associated with a decrease in volume and would be expected to result in traces of compression at the surface.

2.3.5 Partial Melting

Further complications arise from the influence of contaminants such as salts within the ice, and, especially at Saturn and beyond where they should have been able to condense from the solar nebula, volatiles such as ammonia and methanol. These can form compounds with water, and a frozen mixture will contain crystals mostly of pure water-ice but intergrown with crystals of hydrated salts and exotic volatile-rich ices such as ammonia monohydrate ($NH_3 \cdot H_2O$), ammonia dihydrate ($NH_3 \cdot 2H_2O$), or methanol monoammoniate ($CH_3OH \cdot NH_3$). The abundances of these contaminants are debatable, and how their presence influences the strength of ice, and hence its ability to convect, is not well understood, although it is known that the ammonia hydrates each have at least one high-pressure polymorph.

Contaminants also lower the melting temperature of the ice. The likely salts (sulfates, carbonates, and perhaps chlorides of magnesium, sodium, and potassium that would get into the ice through chemical reactions between water and rock) can reduce the melting temperature by only a few degrees, but volatiles have a much greater effect. For example, if there is any ammonia present in the ice, it will be expelled to form a liquid of approximate composition $NH_3 \cdot 2H_2O$ as soon as the temperature rises above 176 K, leaving a residue of pure water-ice. This process is called partial melting (because only a fraction of the solid actually melts), and in this case it happens nearly 100 degrees below the melting point of pure water-ice. A phase diagram to illustrate this effect is given in figure 2.6. At 176 K the ice would stop melting once all the ammonia had been used up. The melt that had been formed would be free either to separate out and migrate independently, or to act as a lubricant between the ice grains, thus greatly reducing the effective viscosity of the ice. Additional contaminants would allow melting at yet lower temperatures; for example, traces of methanol in an ammonia–water mixture would allow the first melt to be produced at 153 K.

Many of the issues raised so far in this chapter are discussed further in chapters 4–7, taking advantage of the observational data that *Voyager* and *Galileo* have provided. In the meantime, we conclude with a synopsis of the outer planet satellites as they were understood before the *Voyager* encounters.

2.4 THE PRE-*VOYAGER* VIEW

In the final years leading up to the *Voyager* missions, the observational evidence and the theoretical modeling discussed in this chapter led to a consensus rock-plus-ice model for the composition of the outer planet satellites, but there was debate about the degree of differentiation that they were likely to have experienced.

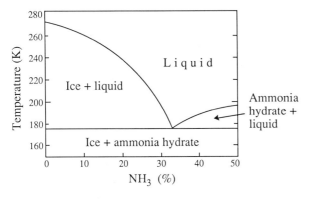

Figure 2.6. Simplified and schematic phase diagram for a mixture of water and ammonia at low pressures. Water-ice and ammonia monohydrate occur as a mixture of crystals at temperatures below 176 K. If the temperature rises above this, for a water-rich mixture (< 30 percent ammonia), an ammonia-rich melt forms with the approximate composition $NH_3 \cdot 2H_2O$, leaving a residue of water-ice as the solid phase.

Figure 2.7. The best image of Ganymede recorded by *Pioneer-10* in 1974. Gross variations in surface brightness can be seen, but no details are visible.

Even if accretion had been homogeneous, it would be possible for a satellite to have a rocky core surrounded by ice if accretion, radioactivity, or tidal dissipation had generated sufficient heat. However, in the absence of any external heat sources, these bodies would have become solid throughout, leaving little scope for any kind of geological activity throughout most of their histories.

Debates of this sort could have gone on indefinitely, and in a fairly futile manner, without the *Voyager* project. Almost everybody was surprised by the variety in the surface histories of the outer planet satellites that was revealed on the *Voyager* images. This chapter has carried the analysis of the interiors of these bodies quite a long way, but it must be remembered that this discussion is based on physical and chemical models with limited observational support, mainly from crude spectroscopic data. Pictures of satellite surfaces played no role in the development of the basic understanding of their evolution as expressed so far, and we have done no more than touch upon the action of geological processes in shaping the faces of these worlds.

It is often overlooked that the *Voyager* spacecraft were not the first missions to the outer solar system. Back in the 1960s there was serious concern that there would be so much small debris in the asteroid belt beyond the orbit of Mars that any spacecraft attempting to pass through this region would be almost certain to collide with something and be destroyed. The pessimistic view went on to state that even if a spacecraft survived passage through the asteroid belt it would be disabled before it came close enough to Jupiter to be of any use because of the intense radiation concentrated by the planet's magnetic field. To see if it was feasible to explore the outer solar system, two spacecraft, *Pioneer-10* and *Pioneer-11*, were launched in 1972 and 1973, respectively. They were small by subsequent standards, only 258 kg each, of which the scientific instruments made up just 25 kg. Much to everyone's relief, both of these spacecraft survived to give the first close-up pictures of Jupiter's cloud formations and, more important for the future missions, reassuring data on the intensity of interplanetary particles and fields. After passing Jupiter, *Pioneer-11* went on to become the first mission to fly through the Saturn system, and after their planetary encounters both *Pioneers* continued to provide data on the interplanetary environment well into the 1990s.

The imaging capability of the *Pioneers* was crude, and little priority could be devoted to obtaining pictures of the planetary satellites. Such pictures as were obtained were undoubtedly sharper than any view seen from the ground, but the sort of detail that they showed did little more than confirm the vague notion of streaky markings on Ganymede (fig. 2.7). Nevertheless, these two pathfinder missions proved that it could be done, and the scene was set for the more ambitious successor missions that are described in chapter 3.

3   *Voyager* and *Galileo*

. . . far outside the flaming walls of the world . . .
Lucretius,
De Rerum Natura

It is not easy to get to Jupiter and even harder to reach the planets beyond. Apart from the hazards posed by the risk of collision and radiation damage, the distances are very great and the journey times involved are so long that the spacecraft must be relied upon to continue to function and stay in radio communication with Earth for several years. As table 3.1 shows, Saturn is about twice as far away as Jupiter, and Uranus is twice as far again.

It took *Pioneer-10* almost two years to reach Jupiter. Bearing in mind that a spacecraft has to fight the Sun's gravity all the way, at the same initial speed it would normally take considerably more than four times as long to reach Uranus. This is a long time to wait for results, especially as a mission is typically in the prelaunch planning stage for several years beforehand. To send a spacecraft on such a long journey to a single distant planet also entails significant risk that it will break down on the way, making the whole mission virtually a washout. By a stroke of immense good fortune, the *Voyager* spacecraft were able to minimize these problems by visiting several planets in turn while still reaching their final targets faster than if they had been sent by a more direct route.

3.1 THE GRAND TOUR

Table 3.1 shows that each planet takes a different length of time to orbit the Sun. This means that their relative alignments are continually changing, as the inner planets overhaul those orbiting beyond them. By a lucky coincidence, during the 1970s the outer planets were positioned such that a spacecraft sent from Earth to Jupiter could carry on to visit Saturn, Uranus, and then either Neptune or Pluto in turn without needing impractically large amounts of fuel to change

Table 3.1 The scale of the solar system, showing the distances of Earth and the outer planets from the Sun. The values quoted are the length of the semimajor axis of the orbit. AU stands for astronomical unit, which is defined by the size of Earth's orbit.

Planet	Mean distance from Sun		Revolution period (years)
	$(10^6$ km)	(AU)	
Earth	150	1.00	1.00
Jupiter	778	5.20	11.86
Saturn	1426	9.54	29.46
Uranus	2868	19.18	84.07
Neptune	4494	30.06	164.82
Pluto	5900	39.44	248.6

course. This meant that even if a Neptune-bound mission were put out of action halfway through its journey, it could already have achieved useful goals at the less-distant planets. In 1966, Gary Flandro, at that time a graduate student working at the NASA Jet Propulsion Laboratory, realized that a spacecraft could tour the outer solar system in this way by using the gravitational field of each planet it reached to accelerate and speed it toward the next target on a gravity-assisted trajectory, more graphically known as the "gravitational sling-shot" effect. Thus was born the concept of the Grand Tour, in which a single spacecraft would take in most of the sights of the outer solar system, needing about a decade to complete the mission. Such a planetary alignment recurs only once in every 180 years, so it is particularly fortunate that space technology was sufficiently well developed in time to take advantage of it. In the event, only the United States, through the agency of NASA, was able to mount an effort along these lines.

NASA's original plan was for a Grand Tour fleet of four spacecraft that would swing past Jupiter and then, among them, to all the planets beyond, including Pluto. In the heady days early in the *Apollo* Moon-landing era the U.S. Congress was willing to allocate NASA the funds needed to equip and launch these missions. However, NASA decided that its future priorities would be the manned spaceflight program, and in particular the development of the Space Shuttle, with the result that the Grand Tour concept was officially abandoned. A much trimmed down version was kept on the books in the form of the *Mariner* Jupiter–Saturn project. This was funded from 1 July 1972 onward, and the team was instructed to design and build two spacecraft that would have a fair chance of surviving encounters with both Jupiter and Saturn and to include an investigation of Titan. As NASA was not willing to commit its budget beyond the projected date of the Saturn encounters, there was to be no instrumentation or electronics specifically designed for targets of opportunity beyond Saturn.

The project was subsequently renamed *Voyager*, and the spacecraft that were eventually built and launched became *Voyager-1* and *Voyager-2*. The nominal mission plan remained the same, and it was not until *Voyager-2* had passed Saturn (and executed a gravity-assist maneuver to speed it toward Uranus) that

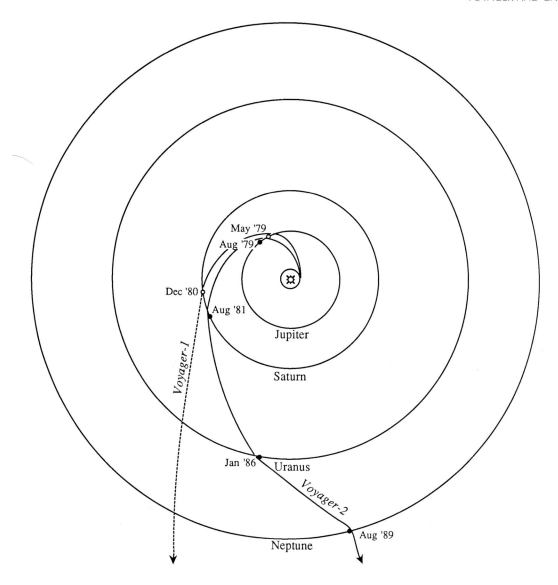

funding for the extended project was approved. The trajectories of the two space-craft are shown in figure 3.1.

3.2 THE *VOYAGER* PROJECT

The *Voyager* spacecraft had several objectives. Foremost among these was gathering information about the masses, magnetic fields, composition, and atmospheric features of the outer planets. As conceived in the mission plans, the satellites were to be of secondary importance. In addition, the *Voyagers* were equipped to measure charged particles and magnetic fields in the long voids between plan-

Figure 3.1. Trajectories of the two *Voyager* spacecraft. *Voyager-1*'s encounter with Saturn flung it onward above the plane of the solar system. After Neptune, *Voyager-2* continued on a course heading below the plane of the solar system.

Figure 3.2. Launch of *Voyager-1* by a Titan-Centaur rocket from Cape Canaveral on 5 September 1977.

ets. The two *Voyagers* were launched in 1977 (fig. 3.2), *Voyager-2* on 20 August and *Voyager-1*, on a faster trajectory that allowed it to reach Jupiter before its twin, on 5 September.

3.2.1 The *Voyager* Spacecraft

The two *Voyager* spacecraft were identical, each weighing 825 kg and carrying 105 kg of instruments for eleven scientific investigations. The basic layout is shown in figure 3.3. Spacecraft designed to operate in the inner solar system usually draw their power from arrays of solar cells that convert sunlight into electricity. This was not a viable option for the *Voyagers*, because sunlight is too weak in the outer solar system; at Jupiter it is less than 4 percent and at Neptune only 0.1 percent of its brightness at Earth. Instead, the *Voyagers* were nuclear powered, using the crude but effective technique of generating electricity from the heat produced in radioisotope thermoelectric generators, which were mounted on a boom on the side of the spacecraft opposite from the sensitive scientific packages.

The cameras and other optical devices were mounted on a stabilized scan platform, which was designed to be fully pointable, allowing a camera to look at its target and yet keep the high-gain antenna aimed exactly at Earth. In addition to being able to point the cameras at a succession of targets while maintaining a radio link with Earth, the movement of the scan platform provided a means of keeping a camera pointing precisely at its target even when the rate of relative

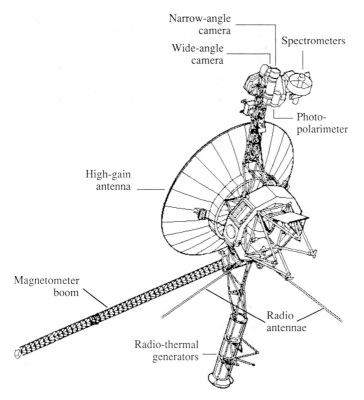

Figure 3.3. Configuration of a *Voyager* space-craft. Hydrazine thrusters for attitude control and trajectory correction were arranged around the ten-sided central mounting. The high-gain antenna was about four meters across.

motion between the spacecraft and target would otherwise have smeared the image. Image-motion compensation of this sort is particularly important in the outer solar system, where the speed of the spacecraft is greater and the level of sunlight is less, necessitating exposures of longer duration (typically 15 s at Neptune).

Near the end of its Saturn encounter, *Voyager-2* suffered a mishap when one of the axes of the scan platform jammed during a preprogrammed slew, leading to an automatic power shutdown of the system. As soon as the problem was realized back on Earth, a fresh command was sent to the spacecraft to turn the scanner power back on, but the jammed axis refused to budge. This could have seriously prejudiced the imaging program at Uranus and Neptune, but in one of the *Voyager* program's many triumphs of engineering compromise, the ground crew managed to free the axis by repeatedly warming it (using a heater intended to keep the lubricating oil at a suitable temperature) and then allowing it to cool again. This eventually freed the axis, but from then on high slew rates were banned by the science team and even medium-rate slews were limited to essential maneuvers necessary to redirect the cameras to new targets. In the end, the image-motion compensation needed to obtain sharp pictures during the Uranus and Neptune encounters was achieved by swinging the spacecraft as a whole, using the attitude control jets. Fortunately there was plenty of fuel left for these, because the original navigation was so good that very little fuel had been needed for course corrections. An additional use of the attitude control jets, introduced at Uranus and refined for the Neptune encounter, was to fire them in a specific

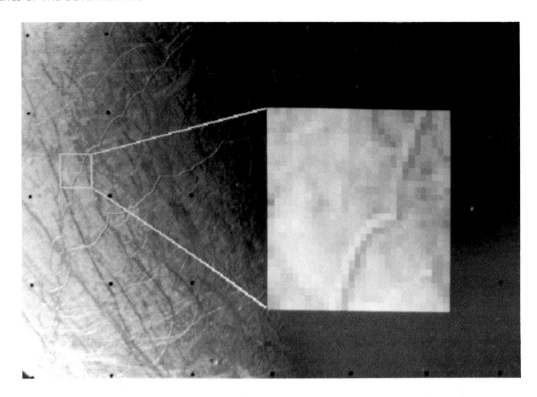

Figure 3.4. Part of a *Voyager* image of Europa, with part of it enlarged to show the individual pixels (picture elements) from which it is constructed. The black dots in the main image are reseau marks (see text for explanation).

combination to compensate for the slight torque caused every time the tape recorder was turned on or off, thus removing an additional source of image smear and pointing inaccuracy.

Each *Voyager* spacecraft carried two cameras: a narrow-angle camera with a field of view just under half a degree across and a wide-angle camera with a field of view of 3.2°. Almost all the *Voyager* pictures in this book were taken with the narrow-angle camera. The wide-angle camera was used more for searching for unknown satellites, imaging planets when they were very close, and obtaining navigational images of star fields. Both cameras were vidicon devices, recording their image in a manner similar to a television camera. However, unlike television, the image was broken down into rows of discrete picture elements, or pixels, before being transmitted. Each image consisted of 800 rows of 800 pixels each, and the intensity of light recorded in each pixel could be anywhere in the range 0–255, a choice of 256 possible levels. Note that this number is 2 to the power of 8, or 2^8, which in computer terms is 8 bits of information. A number in this range is convenient to transmit as a string of code and to handle in an image-processing computer where the image is reconstructed and displayed. An enlargement of part of a *Voyager* image, showing the individual pixels, is reproduced in figure 3.4.

Digital images such as these can almost always be improved by image-processing techniques. Geometric distortions due to the camera can be compensated for by reference to calibration dots (reseau marks) that were etched onto the vidicon faceplate at precisely measured locations. Reseau marks and image defects such as bad scan-lines can be removed and replaced with synthetic pixels or synthetic lines by interpolation from the surrounding area. Contrast can be adjusted as appropriate for dark or bright regions of the surface, and fine details can be enhanced by spatial filtering techniques such as edge enhancement. The image can also be warped to fit a standard map projection if desired. Most of the *Voyager* images reproduced in this book have had the reseau marks removed, all have undergone contrast stretching, and some have been filtered to enhance certain details.

A vidicon is essentially a black-and-white imaging device, because it has no color sensitivity. Color, however, can be an important clue to the composition of surfaces and atmospheres, and so each camera included a filter wheel with a choice of colored filters to accept light of specific wavelengths. The *Voyager* narrow-angle camera had six different filters: ultraviolet, violet, blue, green, orange, and clear (allowing all wavelengths through). To obtain a color picture, *Voyager* had to record separate images in each of three colors, which could then be combined back on Earth to give a colored product. By combining orange, green, and blue the colors appear fairly natural. However, because of the time necessary to rotate the filter wheel to a new position, the viewing geometry had usually changed noticeably between frames, requiring two of the images to be warped to fit the third by geometric image-processing techniques.

As a *Voyager* probe traveled farther away, the strength of its signals received on Earth became weaker, making it harder to transmit its images and other data without an unacceptable level of noise. While at Jupiter the transmission rate was 115,200 bits per second, but this fell to a maximum of 21,600 bits per second at Uranus and Neptune, despite increasing the size of the antennas used in NASA's Deep Space Network and electronically roping in some radio telescopes to help reduce the noise. An image of 800 lines with 800 pixels each consists of 5,120,000 bits, so it would take about four minutes simply to transmit an image from Neptune. At that rate, bearing in mind all the other experiments going on during a planetary encounter, the on-board tape recorders would have filled up before all the desired images could have been handled. Another of the *Voyager* project's technical triumphs was the reprogramming of *Voyager-2*'s on-board computer while it was traveling between Saturn and Uranus, which allowed it to compress the image data. Instead of transmitting an 8-bit number representing the brightness of each pixel, what was sent was simply the difference in brightness between successive pixels. This reduced the average number of bits to four per pixel, representing a reduction of the data volume by a factor of two without loss of information, except in particularly complex scenes where the allocation of bits per scan-line was sometimes used up, leaving blanks that had to be interpolated by image enhancement techniques after reception.

3.2.2 The Encounters

In devising the *Voyager* encounters with the planets and their satellite systems, the mission planners had a near-impossible task. The main requirement was that the encounter should give the spacecraft the right gravitational boost to send it exactly on its way toward its next encounter. Among the other requirements were that the spacecraft should pass as close as possible to the planet and as many of its satellites as could be managed (to enable high-resolution imaging and

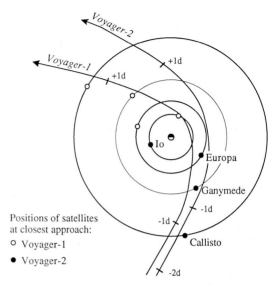

Figure 3.5. *Voyager-1* and *Voyager-2* trajectories through the Jupiter system. Jupiter itself is shown to scale, but the sizes of its satellites have been exaggerated. The tick marks show the time (in days) before and after closest approach to Jupiter. *Voyager-1*'s closest satellite encounter was at a distance of 18,640 km from Io, and *Voyager-2* passed within 60,000 km of Ganymede.

other measurements), and yet avoid flying through any known rings. There was also a requirement to pass behind each planet as seen from Earth (so the attenuation of the radio signal from the spacecraft could be used to determine the structure of the planet's atmosphere), and also to pass behind any rings for similar radio occultation measurements.

During an encounter the spacecraft velocity was so high that there was no scope for slowing down or altering course by using the on-board thrusters, so ideally an encounter would be timed such that the satellites were arranged to let the spacecraft fly past each one in turn, rather like a miniature Grand Tour. In practice this could not be fully achieved, although between them the two *Voyagers* were able to cover the Jupiter and Saturn systems fairly well. As figure 3.5 shows, *Voyager-1* was able to pass close by Io, Ganymede, and Callisto but not Europa, whereas *Voyager-2* passed by Callisto, Ganymede, and Europa but had no opportunity to study Io closely. Similar trade-offs were made at Saturn, but less satisfactorily, as there were more satellites to take into consideration.

Passing close to a satellite is of limited use unless it is possible to obtain good-quality images of most of its surface. How well the two *Voyagers* were able to cover the Jupiter system is indicated in figure 3.6, which shows the distribution of the most detailed images available. Fine-resolution images cover only restricted areas, and Europa in particular had large tracts that were not seen in detail. The coverage of satellites in the Saturn system is patchy in the same way.

The requirement that *Voyager-1* should fly behind Titan (for radio occultation measurements of its atmosphere) meant that it had to proceed onward over Saturn's south polar region, where the gravitational slingshot effect swung it northward and directed it away at an angle of 38° above the plane of the solar system. This precluded any further planetary encounters, so only *Voyager-2* went on to visit the outermost planets.

There was not much choice about *Voyager-2*'s encounter with Uranus; at the time of the encounter the axis of Uranus and its satellite system was pointing

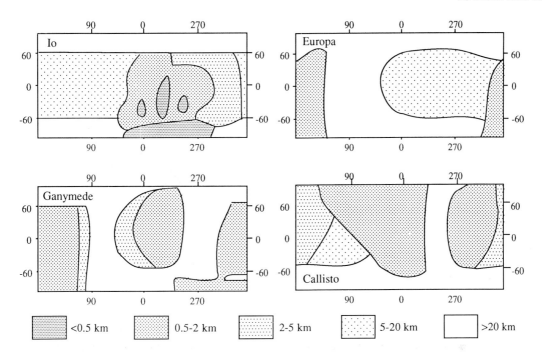

Figure 3.6. Mercator maps showing the distribution of the most detailed *Voyager* images of the galilean satellites. The resolution is expressed as the size of the smallest pixels in the images.

more-or-less directly toward the direction from which *Voyager-2* approached. This meant that *Voyager-2* had to be aimed close to Uranus, rather like a dart at a bull's-eye (fig. 3.7a), with no opportunity to pass close to its outlying satellites on the way in or out. Fortunately, Miranda, the one satellite for which it was possible to arrange a close pass because it was the innermost of the known satellites, turned out to be unexpectedly fascinating. Out at Neptune the situation was different again; there was no onward targeting constraint, so *Voyager-2* was sent as close to both Neptune and Triton as possible, passing into occultation by these bodies at optimum times for signal reception on Earth (fig. 3.7b).

3.2.3 Onward into the Abyss

Voyager-2's encounter with the Neptune system in August 1989 flung the spacecraft onward at an angle of 48° south of the plane of the solar system. After recording a final imaging mozaic sequence showing a view of the whole solar system from outside, the imaging system and other planetary experiments were turned off. However, the cosmic ray, charged-particle, plasma, and magnetometer experiments were left running, with the antenna transmitting on low power, as were similar instruments on *Voyager-1*. Thus began the rather grandiosely named *Voyager Interstellar Mission*, intended to collect data on the strength and orientation of the Sun's magnetic field, the composition and energy of solar wind particles and cosmic rays, and the radio emissions thought to originate at the heliopause marking the boundary between the Sun's magnetic field and the incoming stellar wind.

By 1999 *Voyager-1* was over 72 AU from the Sun and *Voyager-2* nearly 57 AU, both well ahead of the slower-moving *Pioneer* spacecraft with which com-

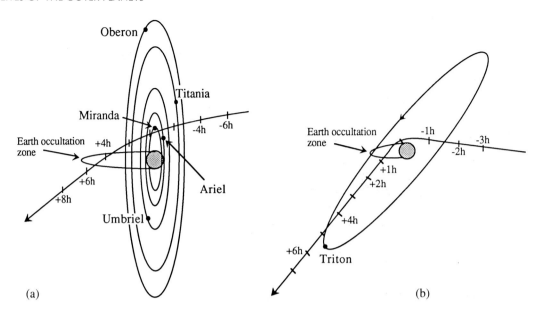

(a)   (b)

Figure 3.7. *Voyager-2* trajectories through the Uranus (*a*) and Neptune (*b*) systems. The tick marks show the time (in hours) before and after closest approach to the planet.

munications ceased in 1997. The original 470 watts of power from each *Voyager*'s radioisotope thermoelectric generator had dropped to about 330 watts, but it was thought likely that sufficient power would remain to keep some instruments operational until at least 2020. *Voyager-1*, which is traveling in the general direction of the Sun's motion relative to nearby stars at a speed of about 3.5 AU per year, will probably reach the solar wind termination shock, a region where the solar wind slows from supersonic to subsonic, in about 2000–2001. It is expected to take several years to traverse this region before crossing the heliopause and becoming humankind's first probe to penetrate the interstellar medium.

3.3 THE *GALILEO* PROJECT

Galileo, named in honor of the discoverer of Jupiter's main satellites, was the first probe to follow the trail blazed by the *Voyagers*. Its goal was to orbit Jupiter to make more detailed and prolonged studies of the planet and its satellites than could be achieved by a flyby mission, and to drop a probe into Jupiter's atmosphere. The project was much delayed by development problems and then the suspension of Space Shuttle operations following the accident that destroyed the Space Shuttle *Challenger* in 1986. *Galileo* was eventually sent on its way from the cargo bay of a Space Shuttle in October 1989. It was boosted out of Earth orbit by a low-thrust upper stage rocket in place of the more powerful type of rocket that would have been used before the *Challenger* disaster but was now ruled to be too hazardous a cargo. As a result, the only way of giving *Galileo* sufficient velocity to reach Jupiter was to use gravitational assists, analogous to the gravitational slingshot effect that accelerated the *Voyager* probes past the giant planets. *Galileo*'s trajectory took it first past Venus (at an altitude of 19,000 km)

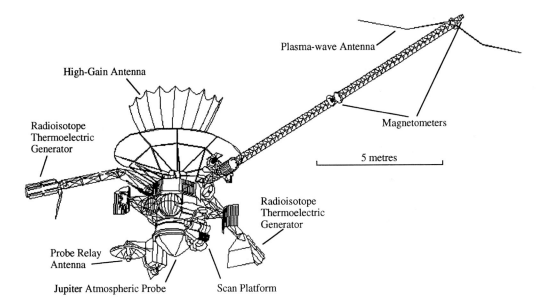

and then on to even lower altitude swings past Earth in December 1990 (960 km) and December 1992 (300 km), between which it made the first ever flyby of an asteroid, 951 Gaspra in October 1991. After its second swing past Earth, *Galileo* was at last on its way to Jupiter. It flew past asteroid 243 Ida and discovered its tiny satellite Dactyl in August 1993, and by good fortune was well placed to obtain direct, though distant, views of the collisions of the fragments of comet Shoemaker-Levy 9 with Jupiter in July 1994. It reached Jupiter in December 1995, becoming the first spacecraft to go into orbit around a giant planet.

At 2223 kg the *Galileo* orbiter was considerably larger than the *Voyagers* and carried a sophisticated 112 kg instrument package (fig. 3.8). Its high-resolution solid-state imaging system was more advanced (using an 800 by 800 array of charge-coupled device detectors with an optical system modified from a *Voyager* narrow-angle camera), and there was a low-resolution near-infrared mapping spectrometer operating between 0.7 and 5.2 μm wavelength designed to help in determining the surface and atmospheric compositions. There were several instruments to probe the magnetic environment and to characterize the dust and charged particles, including those arising from volcanic activity on Io.

Already a triumph by the time it reached Jupiter, *Galileo* had come close to being a disaster because its high-gain parabolic antenna intended to transmit data at more than 100,000 bits per second failed to unfurl despite repeated attempts to unjam it. This meant that all communications had to go via *Galileo*'s small rodlike low-gain antenna, which had a much lower capacity of only 160 bits per second. The reduced rate at which data, particularly images, could transmitted to Earth combined with *Galileo*'s limited on-board storage capacity meant that many intended images had to be scrapped. However, thanks to revised data-compression routines and increased use of multiple receivers of the Deep Space Network in order to pick out *Galileo*'s weak signal from the backgound noise, most of the science and many of the intended images were rescued.

Figure 3.8. Schematic diagram of *Galileo* as it would have looked before the Jupiter atmospheric entry probe was detached. The high-gain antenna is shown incompletely deployed, and was useless. The low-gain antenna, which had to be used instead, is hidden within the bowl of the high-gain antenna.

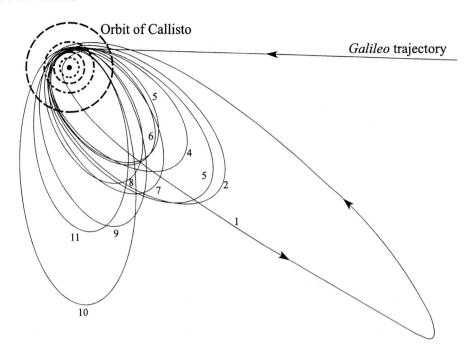

Figure 3.9. *Galileo*'s 1996–97 eleven-orbit primary mission, relative to the orbits of the galilean satellites. Orbits 2, 3, 7, and 8 provided close encounters with Ganymede, orbits 4, 9, and 10 with Callisto, and orbits 5, 6, and 11 with Europa. The closest was during orbit 3, passing only 262 km above Ganymede. The orbital pattern was an economical one: after the initial insertion burn at the start the first orbit, gravitational slingshots during satellite encounters provided most of the change in velocity required to direct *Galileo* toward its next encounter. The *Galileo Europa Mission* contained eight more orbits with close encounters with Europa, followed by four more passing close to Callisto in order to lower the orbiter's perijove (the nearest point to Jupiter) close to the orbit of Io.

Upon arrival at Jupiter, the *Galileo* orbiter made a close pass of the planet, within the orbit of Io, to allow itself to become captured into orbit around the planet (fig. 3.9). Successive orbits were configured so as to make multiple close passes by Europa, Ganymede, and Callisto, but the dangerous radiation belts around Jupiter meant that no further encounters with Io could be contemplated during the primary mission. As a result, the inbound close pass was the only planned opportunity for detailed imaging of Io. Unfortunately, the on-board tape recorder had developed signs of unreliability a few weeks beforehand and it was decided to run this at low speed only. Top priority was, understandably, given to recording and relaying data from *Galileo*'s Jupiter atmospheric probe, whose entry into the atmosphere coincided with the Io encounter. As a result, none of the close-up pictures of Io, which should have shown details as small as 20 m across, were obtained. However, more distant imaging of Io during the remainder of the mission, with a resolution of 3–20 km, proved excellent for monitoring changes in eruptive activity, as we shall see in chapter 7.

Once in orbit around Jupiter, *Galileo* followed a complex trajectory involving a series of encounters with Europa, Ganymede, and Callisto, often at less than 1000 km. These enabled near-global imaging at about 1 km resolution and showed parts of their surfaces with a resolution as small as 20 m. Gravitational and magnetic measurements made during close passes with the satellites greatly advanced our understanding of their interiors, indicating that Io, Europa, and Ganymede have fully differentiated structures with inner, iron-rich cores whereas Callisto appears to be a partially differentiated ice–rock mixture.

Such was the success of *Galileo*'s primary mission that funding was allocated to continue operations for a further two years beginning in December 1997 as the *Galileo Europa Mission*, appropriately abbreviated as *GEM*. The main objectives of *GEM* were to continue detailed mapping of Europa, arguably the most fascinating of the galilean satellites, by means of eight further close flybys. Data from two of these was lost when unexpected anomalies caused the spacecraft to go into "safing" mode, but most of the goals were achieved. *GEM* was scheduled to conclude with four Callisto encounters during mid-1999 that would lower the orbit and allow one or more suicidal dives into Jupiter's radiation belts at the end of the year to obtain the long-awaited close-up pictures of Io.

3.4 MAPPING THE SATELLITES USING SPACECRAFT IMAGES

The *Voyager* and *Galileo* images are virtually the only data we have that can be used to map the surface features of the satellites of the outer planets (although the Hubble Space Telescope has proved capable of mapping large-scale albedo patterns on Titan and Pluto and detecting changes associated with volcanic activity on Io). For most geological work these images can be used directly; for example, a geologist will be interested in investigating the overlapping or cross-cutting relationships between different landforms to determine the geological history of the satellite, because younger deposits should overlie older units, and this information can be seen on the basic images. However, to fit all the information into a global framework, it is necessary to have a reliable map. Using the images to produce a map to cartographic standards is a long and complex procedure that is beyond the scope of this book. The responsibility of achieving this rests primarily with a U.S.Geological Survey team based in Flagstaff, Arizona. Among the tasks that have to be undertaken are interpreting the scale within each image from a knowledge of the relative positions of the satellite and the spacecraft and the pointing direction of the camera, determining the best-fit spheroid to the satellite, and defining a coordinate system so that the data can be plotted on a standard map projection.

It is possible to make a seamless mosaic of the corrected images and to warp them to fit a map projection using digital image-processing techniques, but until the end of the *Voyager* era a more common approach was to redraw the information in the images, using the skills of experienced airbrush artists. This enabled the best details from several images of a single area to be combined, taking account of features that show up best under different illumination conditions or at different scales. In this way a large part of a satellite could be shown in a consistent fashion, drawn so that the angle of solar illumination appears uniform throughout. A certain amount of artistic license was involved in this process, but experience from Mars, which allowed airbrush maps based on early low-resolution images (from the *Mariner* series) to be compared a few years later with higher resolution *Viking Orbiter* images, shows that the artists could usually be relied upon. An example of a shaded relief airbrush map is shown in figure 3.10.

The coordinate system for a synchronously rotating satellite is simple to define. The equator is, as always, 90° from the poles of rotation, and 0° longitude is defined to run through the center of the planet-facing hemisphere. As a result of the inevitable uncertainties in determining the orientation of the images, the ori-

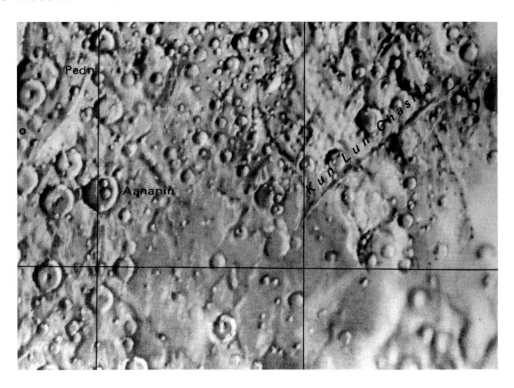

Figure 3.10. Detail from a shaded relief airbrush map of Rhea. The lines are latitude and longitude at 30° intervals. The lack of detail near the southeast corner is because the resolution was poorer, or the view more oblique, on the best images of that area.

gin of the coordinate system is in practice defined relative to the positions of various topographic features that are visible on the images. To refer to these control points, and for general descriptive use, features on the surface have to be named.

3.4.1 The Naming of Names

With several uncharted worlds to cover, the assignment of names is an immense task. To prevent the proliferation of haphazard, inappropriate, and duplicate names, planetary nomenclature is controlled by task groups of the International Astronomical Union. Each of the mappable outer planet satellites has been assigned a theme, usually based on mythology and sometimes connected with the myth from which the satellite's own name was drawn, although the satellite's discoverer is often honored as well. Within the Uranus system there was already a strong theme set up in the names of the satellites themselves: Miranda, Ariel, Titania, and Oberon are all Shakespearean characters (in *The Tempest* and *A Midsummer Night's Dream*), and Umbriel (along with Ariel) appears in Alexander Pope's *The Rape of the Lock*. To continue this theme, the ten smaller satellites of Uranus that *Voyager-2* discovered were each given a Shakespearean name, and the names of surface features on the mappable satellites were also drawn from Shakespeare, except on Ariel and Umbriel, where names of good or bright and evil or dark spirits, respectively, were chosen from mythologies from around the world. In devising names, the International Astronomical Union has tried to avoid any cultural or racial bias throughout the solar system as a whole; for example, the names appearing on the map extract covering Rhea in figure 3.10 are drawn from creation myths of Asian, African, and South American origin. The naming conventions for the satellites discussed in this book are tabulated in the appendix (pp. 213–14).

Craters, such as Aananir on figure 3.10, are given a straightforward name without qualification, but other features are usually given a Latin appendage that describes the general nature of the feature. An example of this sort is Kun Lun Chasma on figure 3.8, where *chasma* (plural: *chasmata*) means a canyonlike feature. Other common terms include *cavus* (plural *cavi*), a hollow or irregular depression; corona (plural *coronae*), an ovoid shaped feature; *facula* (plural *faculae*), a bright spot; *fossa* (plural *fossae*), a long narrow, shallow depression; *linea* (plural *lineae*), an elongate marking; *macula* (plural *maculae*), a dark spot; *mons* (plural *montes*), mountain; *patera* (plural *paterae*), a shallow crater with a scalloped or complex edge; *planitia* (plural *planitiae*), a low-lying plain; *planum* (plural *plana*), a plateau or high plain; *regio* (plural *regiones*), a region; *rupes* (plural *rupes*), a scarp; *sulcus* (plural *sulci*), subparallel furrows and ridges; and *vallis* (plural *valles*), a sinuous linear depression. The use of Latin terms, especially the confusing plural forms, may seem perverse but at least it avoids giving names which imply that we know the true nature and origin of the feature. For example, a possible interpretation of Kun Lun Chasma is that it is a graben, meaning a valley created when parallel faults allowed the floor to subside, but there are other possible explanations such as tidal fracturing, excavation by oblique impact, or volcanic fissuring. Thus, it would be rash (and unkind to future generations) to make its official name the Kun Lun Graben, so we avoid committing ourselves by the use of the purely descriptive term *chasma*. However, this practice does not mean that an effort should not be made to understand the origin and significance of the features observed on the planetary satellites, and it is to that end that most of the remainder of this book is devoted.

4 Icy Lithospheres

... remember of this unstable world ...
Malory,
Le Morte D'Arthur

With such a diversity of worlds among the satellites of the outer planets, you would be excused for supposing that they have too little in common for any mutual thread to run through a discussion of their histories. However, by adopting a geological stance it is possible to set up a unified approach to discussing these worlds, which at the same time helps us to understand the differences between them. To do this, we must look at depth-related changes in the physical properties of the material of which these worlds are made. In particular, changes from "strong-and-rigid" to "weak-and-plastic" states are crucial. By treating chemical composition as of subordinate importance, this approach will also illustrate the close similarities that the icy moons have with Earth and the other terrestrial planets. Since we know most about the Earth, it is useful to start by summarizing the relevant aspects of its internal structure.

4.1 INSIDE THE EARTH

Just about everybody has heard that Earth has a core overlain by a mantle, which in turn is overlain by an outer skin known as the crust (fig. 4.1). The existence of these principal units can be demonstrated by the way that seismic waves generated by earthquakes or explosions are reflected and refracted at the interfaces between them. Subdivisions within the mantle are marked by changes in the speed at which seismic waves travel. The core is recognized as having a fluid outer part because this transmits pressure waves but not shear waves.

The core is distinct from the mantle both physically and chemically. It contains metallic iron and nickel, mixed with about ten percent of a light element that is most likely to be sulfur, potassium, or oxygen. The dense nickel–iron mix-

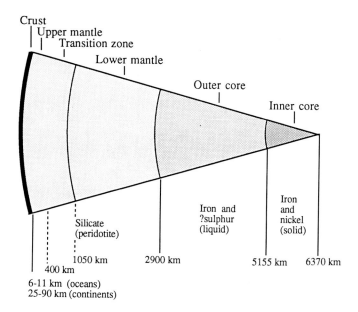

Figure 4.1. The compositional layers within the Earth.

ture is presumed to have segregated gravitationally toward the center after being liberated by melting caused by accretional and radiogenic heating early in Earth's history.

Two of the most common elements in the Earth are silicon and oxygen. An atom of silicon bonds readily with four oxygen atoms to form a tetrahedral-shaped silica unit. This is the fundamental building block of the most common minerals in the crust and mantle. A silicon atom can share one or more of its oxygens with another silicon atom, in which case the silica tetrahedra form chains, sheets, or more complex three-dimensional frameworks. The balance of chemical charges is kept neutral by the incorporation of metallic ions into the structure. In the mantle the most abundant of these are iron and magnesium. The mantle, then, consists essentially of iron and magnesium silicates. It is thought that its chemical composition (essentially that of the rock known as peridotite) does not change significantly with depth and that the divisions within the mantle shown in figure 4.1 correspond to phase changes, where the crystalline structure rearranges to denser forms in a manner analogous to the ice phase changes discussed in chapter 2. Even for Earth, geologists have been able to obtain samples only from the upper part of the mantle, and the exact nature of the deeper phase changes is a matter of informed speculation.

The crust is the bit we live on, and its significance in terms of the planet as a whole tends to be exaggerated. It is, after all, very thin compared with Earth's total size. The crust was formed by partial melts from within the mantle that rose to the surface and solidified. As a result, it is chemically distinct from the mantle, being richer in silica and containing important amounts of elements other than iron and magnesium, which combine with silica to form a new set of minerals. These give the crust a lower density than the mantle, and it is this density contrast that is responsible for the clear seismic division between them, known as the Moho, or Mohorovičič discontinuity.

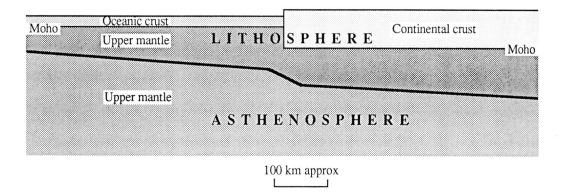

100 km approx

So far as can be told, this description of Earth's layered structure is moderately representative of the other terrestrial planets, especially Venus, whose size and density are much the same as Earth's. There is a complication in the case of the Earth, in that it has two quite distinct types of crust. Oceanic crust is basaltic in composition, that is, with a few percent more silica than in the mantle and with elements such as aluminium and calcium being comparable in abundance to magnesium and iron. Continental crust is more granitic in composition, having even more silica than oceanic crust and with significant concentrations of sodium and potassium, but correspondingly less iron and magnesium. Generally, oceanic crust is about 6–11 km thick, whereas continental crust varies between 25 and 90 km in thickness.

4.1.1 Earth's Lithosphere and Asthenosphere

For studying all except the geochemical aspects of Earth's geological history (and similarly for the other planets), the distinction between crust and mantle is of only secondary importance. To understand the operation of the processes responsible for the deformation of the surface and the volcanism that has resurfaced much of it, we must look instead at how the strength of the planet changes with depth. Mechanically speaking, the uppermost part of Earth's mantle is entirely coupled with the crust. If the crust is displaced, then so is the immediately underlying mantle. Together, the crust and uppermost mantle comprise what is termed the lithosphere, using the Greek root *lithos* (meaning rock). This outermost layer of the solid Earth is rocky in the sense that it can be considered rigid and brittle on most timescales. Its total thickness is 100–200 km below the continents and 100 km or less in the oceans (fig. 4.2). What enables slabs of lithosphere to move around and interact with one another is the fact that directly beneath lies a much weaker layer that flows plastically in response to stress and that will convect if there is an adequate thermal gradient. This layer is termed the asthenosphere, using the Greek root *astheno* (meaning weak).

Earth's asthenosphere is weak because the local pressure–temperature conditions enable the material of which it is made to deform in a near-fluid fashion over geologic time. The asthenosphere does not have a well-defined base, though the mantle is convecting in the solid state all the way down to the core–mantle

Figure 4.2. The most important layers of the Earth in terms of its mechanical properties: the rigid lithosphere and the plastic asthenosphere, showing also the difference between continental and oceanic crust. The mantle below the uppermost part of the upper mantle belongs to the asthenosphere. In most areas the top of the asthenosphere includes a partially molten layer known as the low-velocity zone.

boundary. The situation is compounded by the fact that in many areas of the Earth the top of the asthenosphere contains a small fraction of melt (i.e., it is partially molten), which contributes to the general weakness of the asthenosphere by lubricating the motion between solid grains. This partial melting is another consequence of the pressure–temperature conditions, which encourage certain elements to be squeezed out of the crystals in which they were accommodated and into a liquid phase. If any of this melt reaches the surface, it gives rise to volcanoes. The partially molten zone is recognized by the anomalously low speeds at which it transmits seismic waves, and consequently it is often referred to as the low-velocity zone.

The existence of Earth's asthenosphere was first postulated to account for the "isostatic compensation" of mountain ranges, such as the Himalayas, where the crust is anomalously thick. The crust rides buoyantly on the mantle, so mountain ranges are (almost literally) only the tip of the iceberg, with most of the thickening of the crust occurring below ground level. To allow the crust to project downward into the mantle, there must be a weak layer, the asthenosphere, somewhere near the top of it. This "principle of isostasy" also explains why oceanic crust, being thinner and denser than continental crust, is lower lying. Another demonstration of the existence of a weak layer, which incidentally enables estimates to be made of its strength and rate of flow, is provided by the fact that much of Scandinavia is still rebounding upward at a rate of several millimeters per year in response to the removal of the major ice sheet that covered it during the last glaciation, but which melted away about 15,000 years ago. As we shall see later, similar reasoning can be used to infer the strengths and thicknesses of the lithosphere and asthenosphere on icy moons where major topographic features appear to have undergone isostatic adjustment since their formation.

Equally important on Earth is the role of the asthenosphere in allowing the sideways motion of the lithosphere that occurs in plate tectonic movements. In the 1960s it became recognized that the lithosphere is broken into plates that spread apart at oceanic spreading axes, such as the Mid-Atlantic Ridge and the East Pacific Rise, and are destroyed at subduction zones where the plate motion is convergent and one plate dives down below another, such as beneath the Andes. It used to be thought that these plate motions were driven directly by convection within the asthenospheric mantle, but it is now clear that plate motion is largely independent of this process.

4.1.2 Mantle Convection

Convection of the mantle occurs in the solid state, in the same manner that high-pressure phases of ice will flow, given sufficient time. The rates of flow near the top of the asthenosphere are much the same as the speed of motion of the overlying plates, which approaches 10 cm per year. The speed of convection currents deeper in the mantle may be less, perhaps around 0.1–1 cm per year.

The important difference between the lithosphere and the asthenosphere is that convection can occur in the latter, but never in the former. Most estimates of the viscosity of the lower mantle put it at about only two to three times that of the asthenospheric upper mantle, so for many purposes the asthenosphere can be regarded as including the whole of the mantle below the lithosphere. On a geological timescale it behaves as a fluid, and it has a crucial role in allowing the outward transfer of heat by convection. In contrast, the lithosphere responds to stresses applied from below primarily by fracturing. These stresses may cause the lithosphere to be deformed in a ductile fashion at depth, but it *never* acts as a fluid.

4.1.3 Lithospheres of the Other Terrestrial Planets

The lithospheric thickness of the other terrestrial planets depends on how pressure and temperature increase with depth. The lithosphere on Venus may be thinner than Earth's, as a result of the much higher surface temperature. Its surface has regions that appear to be fold and collision belts, and tracts cut by closely spaced faults or fractures suggestive of fracturing in a thin lithosphere.

The remaining terrestrial planets have much thicker lithospheres than Earth. This is because they are smaller and consequently have lost a greater proportion of their heat since they were formed. For example, Mars has bigger volcanoes than Earth and Venus, but they are much fewer in number, suggesting that on Mars it has been harder to open a pathway for melts to penetrate upward from the partially molten interior. In all provinces but one, volcanic activity ceased about 2.5 billion years ago, and the youngest province appears to have last erupted around 100 million years ago. Taken together with the lack of evidence for plate motions, this points to a lithosphere that was always thick and has grown thicker with time until it is virtually impossible for convection within the sublithospheric mantle to fracture the lithosphere and allow melts to reach the surface.

The Moon is the only body other than Earth for which there are adequate seismic data to detect the internal layers directly. A distinct 60–100 km thick crust forms the top part of a lithosphere that extends to a depth of about 1000 km, about two-thirds of the way to the Moon's center. It seems clear that the Moon's lithosphere has been thickening with time, as a result of cooling, and that it was considerably thinner around 3.4–3.0 billion years ago when most of the major impact basins were flooded by the volcanic eruptions that produced the lunar maria. Mercury probably also has a substantially thick lithosphere today. It bears 3 billion year old scars of an episode of global compression, probably associated with internal phase changes caused by cooling, but shows no signs of deformation or volcanism since then.

Thus, lithospheres and asthenospheres can be recognized in the geological history of all the terrestrial planets. The question is, how can this Earth-derived terminology be applied to the moons of the outer planets, and how can it be used to help us understand their histories? Clearly the chemistry is very different, but as we shall see, the physical principles are the same.

4.2 THE LITHOSPHERE OF AN ICY WORLD

We will leave Io aside until chapter 7; it is effectively a terrestrial planet, in terms of both density and size, so its lithosphere and asthenosphere can be defined as for Earth, although the main source of heat is tidal. For now we will attempt a generic treatment of the icy moons. Variations on this model, as manifested by the differences in their histories, are emphasized in the subsequent chapters.

4.2.1 Compositional Layers in Icy Moons

The first stage is to see if the threefold compositionally based division into crust, mantle, and core can be defined. Most icy moons were probably never hot enough to allow nickel and iron to differentiate out as a core-forming phase (as in Earth), though this does appear to have happened for Ganymede and Europa, which are two of the largest. In all instances, however, we can regard the silicate (rocky) fraction as the main core-forming ingredient, and treat any iron-rich center as simply an inner core.

Referring back to figure 2.3, which shows three stages in the evolution of an icy moon, you should be able to recognize core formation by inward segregation of a dense component in much the same way as was described for Earth, except that this is mostly rock and not nickel–iron. The middle layer (water or ice) would then constitute the mantle, and the outer layer (an undifferentiated ice–rock mixture) would be the crust. The crust in the particular example illustrated is relatively much thicker than Earth's crust, but there is a more fundamental difference in that its origin is due to a completely different process. Earth's crust is a result of chemical differentiation from the mantle, whereas in this icy moon model the crust has the composition of the original, undifferentiated mixture from which the moon formed. Furthermore, the mixture of rock and ice forming this crust will be denser than the icy mantle. The reverse is true on Earth, where the crust is less dense than the mantle (which is why it is rare to find mantle rocks exposed at the surface). However, on an icy moon such as this, if the crust is thin enough and the mantle is weak enough, it should be possible for the crust to fracture and founder, squeezing mantle material, or melts derived from it, up to the surface. Thus, figure 2.3 is a model of an icy moon where it is possible to distinguish core, mantle, and crust, although there are important differences in the origin and relative densities of the crust and mantle compared with the terrestrial planets.

This model of the structure of an icy moon is a fairly middle-of-the-road option; more extreme alternatives were shown in figure 2.5. On the undifferentiated alternative there is clearly no division into core, mantle, and crust, although, as we shall see shortly, this does not preclude the existence of a lithosphere and asthenosphere. On the differentiated model, there is clearly a rocky core, just as in the figure 2.3 example, but because melting in this instance proceeded all the way to the surface, there is no surviving undifferentiated layer, and thus no crust if the definition of the previous paragraph is used. A world like this could be considered to have a core and a mantle only, unless partial melts have risen to the surface to form a thin, but chemically distinct, crust after the fashion of Earth.

The differentiated alternative in figure 2.5 could equally well be the result of homogeneous or heterogeneous accretion, whereas the undifferentiated structure could result only from a homogeneous accretion process. To complicate the range of possible structures still further, there is an important variant of differentiation in a homogeneously accreted moon that must be noted. This is illustrated

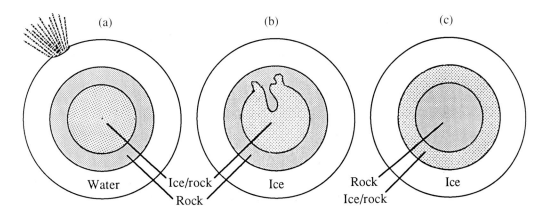

in figure 4.3. In this model the buildup of accretional heat is initially so slow and inward conduction so poor that the center of the body does not melt and remains a mixture of rock and ice. As it grows in size, the amount of accretional heat produced and retained becomes sufficient to melt the ice in the entire outer part of the moon, all the way to the surface. As a result, this outer part differentiates, with the rock segregating downward and the water forming the outer layer. The downwardly segregating rock fragments are brought to a halt when they meet the still-solid rock and ice mixture in the center of the body. The immediate result is therefore a three-layered structure with rock and ice in the center, overlain by rock only, overlain in turn by water only (which freezes to become ice). The rock layer is denser than the rock and ice that it overlies, and so it is gravitationally unstable. If the central rock and ice can be made mobile in any way, perhaps through a small amount of melting, but equally well by solid-state convection, there will be a strong tendency for the overlying rock to swap places with the rock and ice beneath, so the final structure will be a core of rock, overlain by a mantle of rock and ice, with the thick icy shell remaining on the outside.

Some possible arrangements of compositional layers of an icy moon are summarized in figure 4.4. The cores drawn here are just rock, though there could be iron-rich inner cores. What is not possible to show at the scale of the figure is that in option (b), where the ice forms a mantle with no apparent crust, the outermost part of the ice could be greatly enriched in volatiles and/or salts as a result of the upward migration of partial melts. For example, partial melting of a mantle consisting initially of a water-ice and ammonia hydrate mixture (as illustrated by fig. 2.6) would preferentially concentrate ammonia into the melt until the solid residue was pure water-ice. If salts and other volatiles (such as methane, nitrogen, carbon dioxide, and carbon monoxide) were present, these would also be concentrated into the melt. If the melt was expelled upward the result would be a chemically enriched crust overlying a mantle that became purer water-ice as time went on. Such a crust would be more like Earth's present-day crust in origin than any of the other crust options, and is perhaps the likeliest situation for an icy satellite showing evidence of a prolonged history of activity. Moreover, this crust would consist of intergrown crystals of a number of specific compositions (for the simple water and ammonia hydrate mixture these would be pure water-ice

Figure 4.3. Differentiation of a homogeneously accreted icy moon in which melting does not occur at the center. (a) Toward the end of accretion the outer part has melted, allowing the rocky fraction to segregate inward. The central part remains an undifferentiated ice–rock mixture. (b) Gravitational instability causes the denser rocky layer to displace the ice–rock mixture from the center, to form a core of rock surrounded by an ice–rock layer (c), overlain by an outermost icy layer. The time difference between stages (a) and (c) is probably less than half a billion years.

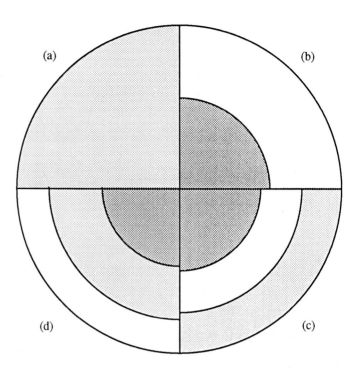

Figure 4.4. Possible compositional layered structures in an icy moon. (*a*) Undifferentiated ice–rock mixture. (*b*) Fully differentiated with a rocky core overlain by ice. (*c*) Partially differentiated with a rocky core, an icy mantle, and an ice–rock crust that has never melted. (*d*) Differentiated, with an ice–rock mixture overlying a rocky core and ice forming the outer layer; this could result from foundering of the crust in (c) or gravitational overturn as shown in figure 4.3. No scale is implied; in fact, (a) is more likely in a smaller moon; (b) in a larger moon, where there may be a separate iron-rich inner core (not illustrated); (c) in a medium-sized moon; and (d) in either a large or a medium-sized moon.

plus hydrated salts and hydrated volatiles such as ammonia monohydrate and ammonia dihydrate) analogous to the mixtures of silicate minerals that make up most rocks on Earth. The first direct evidence of such compositions came when imaging at less than 100 km per pixel by *Galileo*'s near-infrared mapping spectrometer revealed that in certain parts of Europa the normal absorption bands in its spectrum (fig. 2.1) are distorted in a manner consistent presence of hydrated salts such as magnesium sulfate and sodium carbonate.

In addition to internal differentiation, the outermost region of an icy moon that has remained otherwise undisturbed for hundreds of millions of years is likely to have both its physical and chemical nature altered because of the continuing impacts by meteorites and micrometeorites, and exposure to sunlight and charged particles. Impacts vaporize some of the surface ice, and charged particles and solar ultraviolet radiation can break the liberated water molecules into oxygen and hydrogen. The overall process is described as sputtering. The lighter hydrogen escapes rapidly to space, but spectroscopic studies principally by the Hubble Space Telescope and *Galileo* have revealed traces of molecular oxygen (O_2) and in some cases ozone (O_3) on the icy galilean satellites and also on Rhea and Dione, and hydrogen peroxide (H_2O_2) on Europa. Solar radiation affects all longitudes equally, but satellites orbiting within their planet's magnetic influence experience a greater flux of charged particles on their trailing hemispheres.

Although interpreted as evidence for very tenuous oxygen atmospheres in the case of the galilean satellites, most of the oxygen and ozone is likely to be trapped within the ice. *Galileo* near-infrared mapping spectrometer data have also demonstrated absorption features attributable to molecules containing chemical bonds linking carbon to oxygen, carbon to nitrogen, or sulfur to oxygen in the surface ice on Callisto and Ganymede that are probably a result of exposure to charged particles and solar ultraviolet radiation. The long-term end result of this process would be to convert methane and other carbon compounds to a class of organic substances called tholins or even to pure carbon (graphite), which would explain why icy surfaces tend to darken with age and why trailing hemispheres are usually somewhat darker than leading hemispheres.

Some other aspects of meteorite impacts are discussed in the next chapter; essentially, impacts break the surface up into a fragmentary debris known as regolith, and because ice is volatilized more readily than rock, the surface may very gradually become enriched in silicates and carbonaceous matter. However, these are superficial processes, of no great importance in determining global structure.

4.2.2 Ice-Dominated Lithospheres and Asthenospheres

Ask them, and most people will tell you that ice is rigid. Hit it with a hammer and it will shatter. Ask a glaciologist, however, and you will get a different answer. Ice, you will be told, deforms under its own weight, and glaciers can flow downhill at a rate of many meters per year. Clearly, the distinction between rigid and plastic behavior depends on the timescale under consideration. This means that in an icy moon, whether the ice at a certain depth is regarded as belonging to the lithosphere or to the asthenosphere depends on the rate of strain to which the ice is forced to respond. In geological terms, the relevant strain rate is that associated with flow at speeds of the order of 1–10 cm per year, at which the rate of heat transport by convection exceeds that of heat transport by conduction. This lithosphere–asthenosphere transition can be recognized in ice–rock mixtures as well as in purely icy mantles.

So how thick does this make the lithosphere of an icy moon? The ability of ice to deform in the solid state depends very strongly on temperature, so the answer to this question depends on the surface temperature and the thermal gradient (i.e., how quickly the temperature increases with depth). As a rule of thumb, on a geological timescale ice acts in a fluid fashion, flowing by solid-state creep, if its temperature exceeds about six-tenths of the temperature at which it would melt. The equilibrium surface temperature of the moons of Jupiter is about 100 K, dropping to about 80 K out at Saturn. This means that near the surface of an icy moon, water-ice is way below its melting temperature (273 K in the absence of confining pressure) and is well within the field of rigid behavior, even at immensely long timescales. Ice under these conditions is truly rocklike, which is why it is geologists in general and not glaciologists who can profit from the study of icy moons. The low surface temperatures of the icy satellites also explain how steep topographic features, such as crater walls, can retain their pristine form, without slumping downhill as would happen on Earth, where ice is warmer and so able to deform under its own weight much more easily. Figure 4.5 shows in graphical form how the change from fluid to brittle behavior in ice varies with temperature and strain rate.

The temperature inside any planetary body increases with depth, and so the base of the lithosphere in an icy moon is reached at whatever depth the temperature reaches about sixth-tenths of the melting temperature of the ice. As an idea

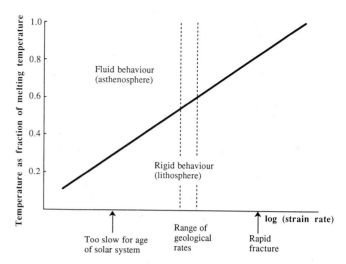

Figure 4.5. Schematic representation of how the temperature at which solid-state creep can occur in ice depends on the rate at which the ice is deformed. Note that the strain rate is on a logarithmic scale. Temperature is expressed as a fraction of the melting temperature, which varies according to pressure (see fig. 2.4).

of scale, on Europa (a dense, massive satellite with a history of significant tidal heating, continuing to the present day) the ice lithosphere is likely to be no more than about 30 km thick, whereas on Callisto (bigger and more massive than Europa, although less dense and lacking tidal heating) the lithosphere today is generally thought to be about 500–1000 km thick. Many of the moons of Saturn, which are considerably smaller, may at present be lithospheric throughout, unless the ice is made more fluid by the presence of contaminants (see below). Because many icy moons have been cooling since their formation, the thicknesses of their lithospheres must have been growing at the same time. We shall see plenty of evidence of this in subsequent chapters.

As we have seen, it is not reasonable to assume that the composition of either the lithosphere or the asthenosphere of an icy moon is pure water-ice, because there are likely to be various contaminants mixed in with the ice. The most common, which is likely to be found in icy moons throughout the solar system, is rock. Fragments ranging from dust size to house size and above may be evenly distributed or concentrated in a particular layer depending on the differentiation history of the moon. Rock fragments dispersed within ice increase its rigidity at a given temperature and so will tend to make the thickness of the lithosphere greater than in a pure ice model. Counterbalancing this is the fact that because rock is denser than ice, the pressure in a rock and ice lithosphere must increase more rapidly with depth than within a pure ice lithosphere. As figure 2.4 shows, the melting temperature of water-ice decreases with pressure up to about 1 kbar (which is equivalent to a depth of about 180 km in Callisto and is not reached at all in moons smaller than Iapetus), so this pressure effect makes the ice itself less rigid at a given temperature. In general terms, then, dispersed rock fragments can make an icy lithosphere either thicker or thinner depending on a complex balance of effects.

Another important contaminant is ammonia (NH_3), which is likely to have been able to condense from the solar nebula in significant amounts at Uranus and beyond, and possibly even at Saturn. Cosmic abundances suggest that the ammonia could make up at most no more than about 20 percent of the total mass of the ice in planetary satellites. The principal effect of mixing ammonia-ice (either ammonia monohydrate or ammonia dihydrate) with water-ice is to make the ice more rigid, by hindering deformation within grains. This is essentially a lithospheric process, and probably does not act to increase the viscosity of the asthenosphere because, as described in chapter 2, ice that is contaminated by ammonia undergoes partial melting to produce an ammonia hydrate liquid of approximate composition $NH_3 \cdot 2H_2O$ if the temperature rises above 176 K. This leaves a residue of pure water-ice with an intergranular fluid of the ammonia hydrate melt. If this melt remains at depth, as an intergranular fluid, then it would be analogous to the partially molten low-velocity zone within Earth's asthenosphere. Salts, likely to be present in all icy satellites, have a much more minor effect in melting point depression. Like rock fragments, salt crystals probably increase the viscosity of ice within which they sit.

At Uranus and Neptune there may also be other contaminants in the ice, notably methane (CH_4), carbon monoxide (CO), or nitrogen (N_2). For every six molecules of water in ice I (but not the higher pressure forms), there is a space within the crystalline lattice large enough to hold a molecule of any of these, in a structure known as a clathrate. This will have a poorly understood effect on the rigidity of the ice, but it seems clear that if the temperature rises sufficiently to drive the contaminant out of its lattice site and into a liquid phase, even if the contaminant occurs only in small amounts, this intergranular fluid must greatly reduce the effective viscosity of the ice as a whole, thus driving the lithosphere–asthenosphere boundary closer to the surface. Another effect of caging volatile molecules in a clathrate is to reduce its thermal conductivity to about a twentieth of that of pure water-ice. This would make it hard for heat to escape through a clathrate lithosphere, and so internal temperatures even from radiogenic heating alone could become sufficiently high for partial melting to begin.

4.2.3 Icy Volcanism

We have seen that icy moons can be layered in both compositional and mechanical senses. The ice in the asthenosphere is capable of flow at rates sufficient to sustain convection. What about the production of fluids that could behave in a similar way to molten rock on Earth? In other words, are "igneous" processes possible?

We have already remarked that although dominated by water-ice, the crusts of icy satellites could contain crystals of hydrated salts and volatiles. If these intergrowths formed by freezing of partial melts, then they are close analogs of igneous rocks on Earth. Doubtless some melts freeze at depth to form intrusions, but others are erupted at the surface. Several examples are discussed in subsequent chapters. Because of the low temperatures involved, eruptions on an icy moon are sometimes described as "cryovolcanic," but this should not be allowed to confuse the essential similarities with volcanic processes on the terrestrial planets. For example, an ammonia-water melt (approximately $NH_3 \cdot 2H_2O$) formed by partial melting of water-ice contaminated with ammonia will probably be slightly less dense than the remaining solid ice and will definitely be less dense than the solid if this contains rock fragments. Such a melt would therefore tend to percolate upward, but of course it would all solidify as soon as it reached a region where its temperature dropped below 176 K. However, if a sufficient amount of such a melt accumulated in a pocket below the lithosphere before migrating up through a crack, en masse and rapidly, it could reach the surface before it froze. This melt would have physical properties (notably viscosity and yield strength) that under the low surface gravity of an icy moon would cause it to behave in a manner similar to a basaltic lava flow on Earth. In contrast, if the melt had undergone a significant amount of crystallization during its ascent, its viscosity and yield strength would be increased into the range exhibited by more silica-rich terrestrial flows, such as dacites and rhyolites. Thus even simple ammonia–water cryovolcanic melts provide scope for a variety of icy volcanic landforms that have terrestrial volcanic analogs, and the scope is further increased in the event that methanol, other volatiles, and salts are also present in the system.

Volcanism in the form of salty water (brine) is less easy to achieve in the absence of volatiles. Salts lower the melting temperature by only a few degrees relative to that of pure water-ice, so higher temperatures are required to yield brines than for ammonia–water mixtures. The temperature required for partial melting would be hard to reach because of the efficiency of solid-state convection. Moreover, any brine produced would be likely to freeze before it reached to the surface. Another problem is that many brines are denser than ice and so would not tend to rise, unless the ice were weighed down by rocky impurities. Any salty magmas reaching the surface of an icy moon are likely to be slushy crystal-rich mixtures or could even be mostly ice mobilized by trace amounts of intergranular volatile fluids.

Another way of transporting the products of melting and mobilization at depth to the surface is by gas-driven processes, comparable with explosive volcanism on Earth. Trace amounts of ammonia in water could vaporize as the pressure decreased near the surface, or the passage of an $NH_3 \cdot 2H_2O$ melt through ice containing methane or some other volatile in clathrate form could cause the clathrate forming the walls of the crack to decompose into water-ice and gas (fig. 4.6). Both these processes would drive the melt to the vent in an explosive fash-

Figure 4.6. A possible thermal gradient within an icy moon, showing the depth at which an $NH_3 \cdot 2H_2O$ melt could form and the path this magma would follow if it were to ascend rapidly without cooling. The solid line shows the conditions under which a methane clathrate decomposes to water-ice and methane gas. Thus the passage of the magma could liberate methane gas from the walls of the crack, resulting in an explosive eruption. The relationship of depth to pressure depends on the size and mass of the satellite concerned; the depth scale here is for Iapetus.

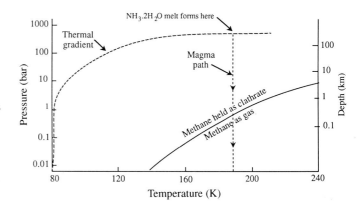

ion, where it could be distributed either as frozen shards in a gas-driven jet or as a frothy liquid.

Now that we have examined the general physical and chemical factors that control the geology of the icy moons, we are in a position to examine these worlds individually in the following chapters, to see what tales they have to tell about the forces that have shaped their surfaces.

5   Dead Worlds

Strike flat the thick rotundity of the world
                                        Shakespeare,
                                              King Lear

Rather than describe the moons of the outer planets one planet at a time, they are presented here in classes defined by the extent to which each exhibits evidence of geological processes. This chapter describes the worlds that show few or no signs of geological activity on their surfaces except those associated with and recorded by impact cratering. As introduced in chapter 1, heavily cratered surfaces must be older than less cratered areas because the latter can be assumed to have once been heavily cratered themselves but to have had their craters obliterated by geological processes such as tectonism (deformation) or volcanism. Chapter 6 discusses worlds exhibiting clear signs of such geological resurfacing having overprinted and partly destroyed the earlier cratered terrains, and where such activity evidently belongs to a past era. Chapter 7 covers those worlds where this sort of geological activity is continuing today or is at least very young. Finally, chapter 8 hazards some predictions about what conditions might be like on the few icy worlds of whose surfaces we have no good images.

The satellites included in this chapter are Callisto (Jupiter); Mimas, Rhea, and Iapetus (Saturn); and Umbriel and Oberon (Uranus). We will start with Callisto because it is the most fully (and longest) studied world in this class. Almost everything we know of the geology of these worlds is derived in one way or another from the study of their craters, and before delving into cratering processes in the outer solar system, it would be best to summarize what is known about the production of craters closer to home.

5.1 IMPACT CRATERING

The planetological importance of understanding cratering first became apparent after studies of the Moon by Eugene Shoemaker and his colleagues in the early 1960s. These studies demonstrated a relationship between the relative age of stratigraphic units on the Moon as deduced from superposition (younger units overlying older ones) and the number of craters per unit area of surface (i.e., the "crater density"). Older units have a denser distribution of craters. Since then, the stratigraphic units have been fitted into an absolute timescale by the radiometric dating of samples returned to Earth by the six *Apollo* manned landings (U.S.A.) and the three *Luna* unmanned sample-return missions (U.S.S.R.) that took place between 1969 and 1976.

Relative dating of the terrain units on the airless world of Mercury and the thinly atmosphered planet Mars can also be performed using cratering statistics. Their crater size distributions resemble those on the Moon, and dynamics arguments suggest that a population of impacting bodies whose orbits brought them into the inner solar system would become fairly evenly scattered as a result of gravitational interactions. Thus, although there is as yet no direct calibration from actual samples, the cratering timescale set up on the Moon is applied with reasonable confidence throughout the inner solar system. However, a variety of evidence suggests that the bodies responsible for the cratering of the moons of the outer planets were not part of the same family, or "population," that bombarded the inner solar system. To elaborate on this, it is necessary to delve into the abstruse realm of crater statistics.

5.1.1 Crater Statistics

The Moon is dominated by two quite distinct terrains: the ancient highlands and the younger basaltic maria that resulted from volcanic flooding of most of the major impact basins. The highlands are extremely densely cratered because they were subjected to bombardment by debris left over from the formation of the solar system. However, most of this had been mopped up by the time the mare surfaces were formed, and these areas consequently have a considerably lower crater density. The contrast between the two terrain types is apparent in figure 5.1.

Simple visual comparisons are not sufficient to extract all the information that is lurking within images such as this. If we want to use craters to date a surface precisely or to tell whether a particular world was bombarded by the same population of impactors as another, it is important to quantify both the numbers and sizes of the craters present. The standard way to do this is to count per unit area of surface the number of craters whose diameters fall into successive size increments. The most straightforward way to display the results of this counting is to plot an "incremental size–frequency distribution curve." On such a graph, the crater diameter is plotted on the horizontal axis and the vertical axis shows the number of craters per square kilometer having a diameter within each incremental range.

An example of size–frequency distribution curves for lunar highlands and lunar maria is shown in figure 5.2. Both the axes are on a logarithmic scale in

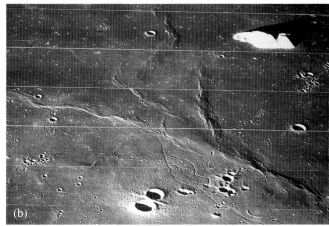

Figure 5.1. Two parts of the Moon's surface photographed by an unmanned *Lunar Orbiter*, shown at similar scales. (*a*) Highland terrain, virtually saturated with impact craters inherited from the late heavy bombardment of the inner solar system. (*b*) A darker mare surface, which, being younger, has far fewer craters. An isolated peak of highland material rises above the lava-flooded mare at the top right.

order to cover a wide range of crater sizes and the vastly greater frequency of smaller craters than larger ones. This graph shows that the lunar highlands have a crater size–frequency distribution similar to that of the maria, except that the highlands have about a hundred times as many craters of a given size per unit area. If the rate of impacts had remained uniform through time, this would mean that the highlands must be about a hundred times older than the surfaces of the maria, but we know from radiometric dating that the highlands are around 4.0 billion years old and that the maria range from about 3.9 to about 3.1 billion years in age. This demonstrates a dramatic decrease in the cratering rate at about 3.9 billion years ago, marking the end of the late heavy bombardment period. Crater densities and radiometric ages on young ejecta blankets on the Moon suggest that the rate of impacting has remained fairly constant since at least 3.5 billion years ago, although there may well have been brief flurries of crater formation at times of enhanced cometary flux.

If a surface had been bombarded by a single population of impactors consisting of few large objects but many small ones (as is implied by the known present distribution of potential impacting bodies in Earth's vicinity), then we should ex-

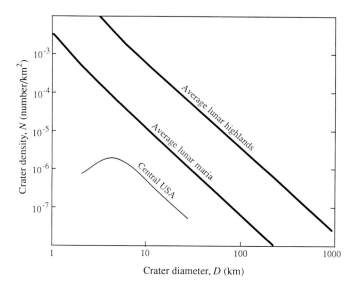

Figure 5.2. Incremental size–frequency distribution curves for the two major terrain divisions on the Moon. There are many more small than large craters, and the lunar highlands have approaching a hundred times more craters than the maria. Variations in crater density within these major units (not illustrated) are used for relative dating. A curve for the central United States is also shown; there are many fewer craters because the surface is much younger, and the curve turns down at small crater sizes because small craters have been removed by erosion.

pect a roughly straight line on an incremental size–frequency plot, as indeed is shown in figure 5.2. The gradient is typically about –2 in the inner solar system, except that it steepens to –3 or –4 for craters smaller than a few kilometers in diameter, many of which are secondary, having been produced by major blocks of ejecta from primary impacts. In practice, the gradient can become artificially less at the small-crater end of the distribution owing to a variety of circumstances, including the small craters being harder to see on the images, their being more easily buried by ejecta from subsequent larger impacts, and the surface becoming saturated with small craters so that for each one created there is, on average, a similar one destroyed. There are also subtle inflexions of the curve when a second population of impactors with a different size–frequency distribution has partly overprinted the record of the earlier bombardment.

These effects are hard to see on an incremental size–frequency plot. The standard technique to overcome this is to plot how far the observed distribution departs from a line with a gradient of –3, using what is known as a "relative size–frequency distribution plot" (fig. 5.3). This has crater diameter on the horizontal axis as before, but shows a parameter R on the vertical axis, which is the ratio of the observed distribution to the function $N = D^{-3}$, where N is the number of craters per square kilometer within the size increment at diameter D. The power –3 has no fundamental physical significance, but it is mathematically convenient. We will refer to relative size–frequency distribution plots when discussing the history of the icy moons. The concept may seem complicated, but to use these plots only two points need to be appreciated. First, the higher the value of R at a given crater diameter, the greater the density of craters. Second, if the crater distribution falls on a horizontal line, then it has a straightforward impact history, but if the line is curved then either there has been more than one population of impactors involved or something has removed craters in part of the size range subsequent to their formation. On a curved line, a segment that slopes upward from left to right indicates relatively

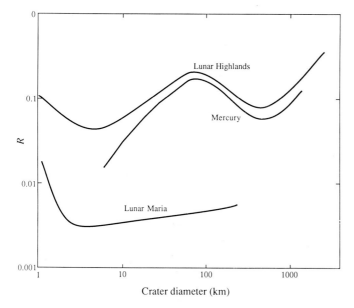

Figure 5.3. Relative size–frequency distribution plot showing the differences in the distribution function for craters on the lunar highlands and the lunar maria. This indicates that the population of impactors that produced most of the highland craters differed from the later population that produced all the impact craters on the maria (which are younger than the highlands). However, there is a clear similarity between the oldest cratered terrain on Mercury and the lunar highlands, showing that they were cratered by the same population of impactors.

more larger craters, whereas a segment that slopes downward indicates relatively fewer larger craters.

Figure 5.3 makes it easy to see that there are actually important differences in the crater distributions between the lunar highlands and the maria. Comparable differences in older and younger cratered areas have been found on Mercury and Mars. It is fairly clear that the postmaria craters represent impacts by asteroids and comets of similar characteristics to the bodies observed today still in orbit. The highlands bear both pre- and postmaria craters. Unless the subtly different crater distribution on the highlands is an artifact of some crater-modification process, it must indicate a different size–frequency distribution for the premaria impacting bodies responsible for the late heavy bombardment.

5.1.2 Crater Formation

The foregoing discussion has made the implicit assumption that the craters we see on other worlds are the result of impacts and not of some other process capable of producing craters, such as volcanism. The origin of lunar craters was controversial for the hundred years or so leading up to the *Apollo* landings, with the proponents of impacts and of volcanism evenly matched. Nowadays the impact theory is very much the established view, although there are still a few heretics around.

Reasons for believing in the impact origin of craters are the closely comparable size–frequency distributions on the Moon, Mercury, and Mars (which indicate a common external control), and the abundance of craters on asteroids (which are bodies too small to have experienced volcanism). Moreover, a buildup of impact craters on an old surface seems inevitable, based on the measured current rate of impacts on Earth's surface. For every 10 million km² (an area the size

Figure 5.4. The Barringer meteor crater, Arizona.
(*a*) View of the 1.2 km diameter raised rim, from the
northwest. (*b*) A rather exaggerated claim displayed on
the approach to the crater, which at least serves to
make the point that the processes by which extraterres-
trial impact craters formed are in little doubt, as a result
of detailed field and laboratory studies on Earth.

of Europe or the United States), a crater larger than 1 km in diameter is formed,
on average, once every quarter of a million years, and one larger than 10 km in
diameter every ten to twenty million years. Bigger impacts are even less frequent
but may have consequences global in scale; for example, the now buried 65 million year old, 170 km diameter Chicxulub crater in Mexico is associated with a
"mass extinction" event at the Cretaceous-Tertiary boundy—an episode of
global devastation when many plant and animal species (including the last of the
dinosaurs) died out.

    The morphology of lunar craters is distinctly unvolcanic (e.g., their floors are
lower than the level of the surrounding terrain), whereas they have all the characteristics of natural impact craters on Earth—the famous Barringer meteor crater

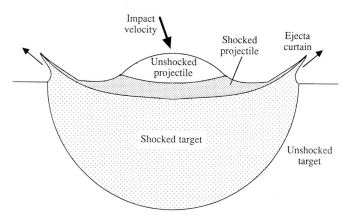

Figure 5.5. The compression stage of an impact, seen in cross section a fraction of a second after arrival of the projectile. Melted and fragmented material derived from the shocked regions of the projectile and target feeds the beginning of a conical ejecta curtain. Material fed into the ejecta curtain at this stage has the highest velocities and will be the last and farthest to be deposited. The geometry is more or less independent of the angle at which the projectile hits the target surface unless the impact is extremely oblique.

in Arizona being perhaps the best known (fig. 5.4)—and of impact craters formed in scaled-down laboratory experiments. This is not to say that there are no volcanic craters on the Moon, but they are far fewer than the impact craters and can be recognized by associated volcanic flow features and other characteristics.

In this book the conventional view will be followed that heavily cratered terrains are the result of a long history of impacts, and that a volcanic crater can (almost always) be recognized on morphological grounds. The processes that occur in an impact are by now fairly well understood. Typical impact velocities in the solar system are of the order of 10 km s^{-1}. At such high velocities, the first thing that happens when a projectile hits the target surface is that shock waves propagate forward into the target and backward into the projectile. For a fraction of a second the pressures are extremely high, far greater than those necessary to fracture the material of which both the target and the projectile are made, and indeed great enough to create high-pressure shocked forms of certain minerals, as has been documented both on Earth and on the Moon. This short interval is known as the compression stage of an impact, during which an upward and outward motion of ejecta consisting of vaporized, melted, and pulverized material from both the projectile and the target forms the beginning of an ejecta curtain (fig. 5.5). By the time the projectile has completely disintegrated, the shocked region extends to about one projectile diameter into the target.

The second stage of the crater-forming process, the excavation stage, continues with the growth of a transient cavity, feeding ever more material into the ejecta curtain (fig. 5.6a). The transient cavity is hemispherical at first, but in a large impact it reaches a critical depth, after which it grows laterally only. It continues to grow until the energy of the impact is used up, and the last material to be supplied to the ejecta curtain joins it at low velocities, barely managing to flop out of the cavity, creating an overturned flap that forms the crater rim (fig. 5.6b). The crater dimensions at the end of this stage are one to two orders of magnitude greater than those of the original projectile, and it takes about 10 s to form a 1 km diameter crater and about 100 s to form a 100 km diameter crater. The crater that remains may have the shape of the transient cavity as it was when the excavation stage petered out, but for craters larger than a threshold size (which depends on gravity and the composition of the target), inward collapse of the walls further enlarges the crater, with the formation of terraces and concentric rings. In

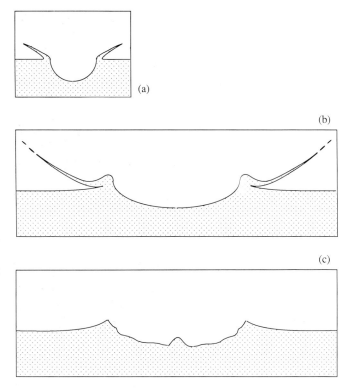

Figure 5.6. Crater formation, seen in cross section: (a) and (b) are during the excavation stage, and (c) is after modification. In (a) the transient cavity is hemispherical. A conical ejecta curtain, consisting of material from both the projectile (now completely destroyed) and the target propagates outward as the transient cavity grows. Stage (b) is several seconds or tens of seconds after (a) and is shown at about half the scale; the last material is entering the base of the ejecta curtain at low velocities and the transient cavity has reached its maximum extent. The upper, outer part of the ejecta curtain continues to spread and fall out ballistically and will give rise to an ejecta blanket around the crater, including secondary craters where large blocks of ejecta hit the surface. If there is no atmosphere, the ejecta curtain as it moves outward becomes increasingly separated into filaments that produce rays and loops of ejecta. In (c) the crater has been enlarged by inward collapse of its walls, and a central peak has formed.

addition, a central peak may form within the crater, either as a result of ballistic rebound or because of push from the slumping walls (fig. 5.6c).

On the Moon there is a progressive change in crater morphology as the size increases. Craters less than 10 km in diameter have simple bowl shapes, whereas essentially all craters larger than about 40 km in diameter have flat floors, central peaks, and terracing on the inner side of their rims. For those craters whose diameters exceed about 300 km it is common to find concentric rings encircling the crater, giving rise to the name "multiringed basin." The largest of these on the Moon have diameters of over 1000 km.

There are similar size-related morphologic transitions in craters on the icy moons. The critical diameters vary from one to another mainly because of their different surface gravities, and there is an additional class, central pit craters, which are not found on rocky worlds, except rarely on Mars, where they may be related to water- or ice-saturated rock. The replacement of the central peak by a pit may be due to the gravitational collapse of a warm, possibly liquid, transient peak. The morphologic series with increasing scale on icy moons may be summarized as follows: bowl-shaped craters < central peak craters < central pit craters < multiringed basins. Examples of these are shown in figure 5.7.

5.1.3 Regolith

A surface that has been subjected to a prolonged period of bombardment loses its original physical constitution. In the case of the Moon, the surface is covered by

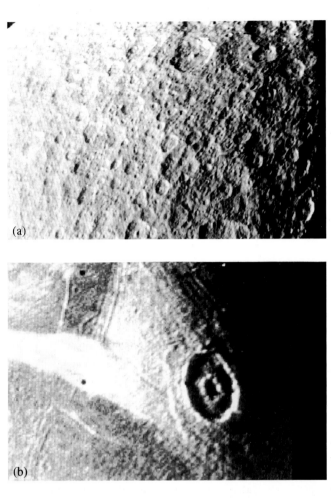

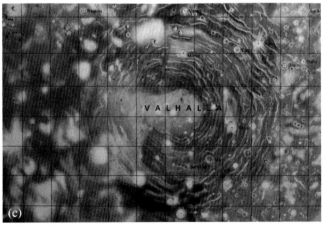

Figure 5.7. The main crater morphologies that occur on icy moons with increasing scale. (*a*) Smaller bowl-shaped craters and larger central peak craters on Rhea. (*b*) A central pit crater on Ganymede. (*c*) An airbrush map showing the 4000 km diameter Valhalla multiringed basin on Callisto.

a "soil" or regolith of fragmented lunar rock and glassy particles produced by impact melting. Less than about one percent of the material present is derived directly from meteorites. The grain size of most of the regolith is less than about a tenth of a millimeter (although there are plenty of larger lumps), and this is thought to result from continual reworking of the surface by micrometeorite impacts as well as by the less frequent stirring up that occurs when a larger impactor strikes.

It is logical to expect that the cratered icy moons have similar regolith-covered surfaces. This is supported by three kinds of observation. The first is photometric study of the way the moons scatter sunlight as a function of incidence angle, which shows that their surfaces are granular, having fine-scale roughness that tends to increase with age. The second is Earth-based polarizing radar observations, which generally suggest a deep regolith of many tens of meters. The third is thermal infrared measurements of the rates of cooling during eclipses, which demonstrate a low thermal conductivity compatible with a fragmental surface layer. It is worthy of note that micrometeorite impacts probably help to concentrate rock dust in the regolith by ejecting water molecules or oxygen and hydrogen ions from the ice, through sputtering, which over long time periods results in a net loss of ice to space.

5.2 IMPACT CRATERING IN THE OUTER SOLAR SYSTEM

Post-late heavy bombardment cratering in the inner solar system is attributed mainly to comets, with a smaller but significant proportion caused by asteroidal fragments. Jupiter lies beyond the main asteroid belt, and it is doubtful whether the supply of asteroidal debris is now or ever has been comparable with that in the inner solar system. The likelihood of the same population of impactors extending to Saturn and beyond is even less. Arguments about variations in the flux of crater-forming impactors throughout the solar system were at least partly resolved by the *Voyager* missions, which provided the first images on which craters could be counted and from which crater size–frequency plots could be drawn. Figure 5.8 shows a relative size–frequency distribution plot, like figure 5.3, with curves for Callisto and several more remote satellites. Without going into details just yet, neither the shapes nor the positions of the peaks of the curves for these bodies match those in the inner solar system. This implies that the size–frequency distribution of the impacting bodies themselves must have been different. There are also significant differences in the curves between Callisto and the moons of Saturn, and even within the family of Saturn's moons.

5.2.1 Crater Scaling Between Satellites

When comparing crater statistics between diverse bodies, it is well to be aware of the pitfalls. An impactor with a given mass and impact velocity will produce a smaller crater on a world with high surface gravity than on one with lower surface gravity. In addition, target material such as ice that has comparatively low density and low strength will form larger craters than a rocky target with

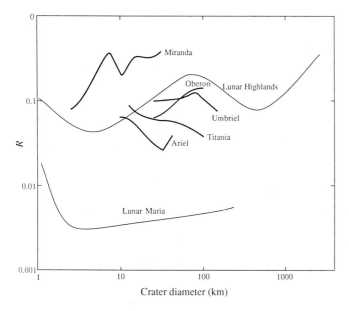

Figure 5.8. Relative size–frequency distribution plot showing the difference in crater populations between the Moon and the most densely cratered regions of several heavily cratered outer planet satellites.

high density and high strength. The predicted values depend on the details of the model used, but in general crater diameters on the icy satellites should be about one and a half to three times greater than on the Moon, for the same impact energy.

The sizes at which craters undergo transitions in morphology also vary, and this is something we can observe directly. The transition from simple bowl shapes to more complex craters with a central peak occurs below 10 km for Callisto and Ganymede but at around 30 km for smaller worlds such as Mimas. Despite their slightly lower surface gravity than the Moon, the morphologic transition on Callisto and Ganymede occurs at a smaller size than it does there, which is evidently a result of the lower strength and density of their icy surface material compared to the Moon's rocky surface.

There are also considerations involving the orbital motion of a synchronously rotating satellite and the gravity field of its planet. Impacts with projectiles originating beyond the satellite system in question should be most common on the leading hemisphere of a satellite, because this is the side that is traveling toward danger. In the case of a satellite in synchronous rotation, one hemisphere is fixed in this role, so it should accumulate several times more craters than the trailing hemisphere if the tidal lock has remained unbroken throughout the satellite's impact history. The effect increases with the orbital velocity of the satellite, so it should be more pronounced on the inner satellites of a planet, which travel the fastest. Externally originating projectiles are subject to another important effect in that incoming projectiles will be deflected toward the planet by its gravitational field, so they are more likely to hit an inner satellite than an outer one. The effect of this gravitational focusing mechanism would be to make the present cratering rate on Mimas (Saturn's innermost major satellite) about twenty times that on Iapetus (its outermost big satellite).

5.2.2 Impactor Populations

Scaling effects appear to be capable of shifting the crater size–frequency distributions for icy moons toward larger diameters by up to a factor of about three, and up and down by an order of magnitude or so. There is no way that this could account for the disparities among the distributions shown in figure 5.8, where there are clearly different slopes. This problem lends support to the concept that different populations of impactors have been involved in each part of the solar system.

The most obvious trait in the crater size–frequency distribution for Callisto is a drop-off at diameters greater than about 60 km. This is repeated in comparable data for Ganymede (not shown in fig. 5.8). It appears, then, that the population of impactors affecting Jupiter's satellites was deficient in impactors sufficiently massive to create large craters. As we shall see shortly, some large craters have probably been lost because they subsided under their own weight when the lithosphere was thin and warm, but it is unrealistic to appeal to this mechanism to account entirely for the loss of the large craters. Thus, most people agree that Callisto and Ganymede record a flux of impactors substantially different from those that affected the inner solar system. An unfortunate corollary of this theory is that we cannot use crater densities at Jupiter for absolute dating, because there is no way to link these crater statistics to independently determined ages from the Moon or Earth.

At Saturn, the situation is even less satisfactory. There are considerable differences in the crater size–frequency distributions even among its heavily cratered moons. No moon of Saturn shows a correlation between the density of craters and position on the globe. This suggests that the tidal locking of the rotation has been broken several times by impacts large enough to impart sufficient angular momentum to increase the rate of spin temporarily, and that each time synchronous rotation was restored (by tidal forces) the moon presented a different face to its planet. Within the Saturn system as a whole, two populations of impactors have been proposed, known as Population I and Population II. Population I is older, and its traces have been partly erased on younger surfaces on Dione and Tethys. It produced a substantial proportion of large craters. Population II yielded a deficiency of large craters and is the dominant population on younger surfaces. In figure 5.8 the curves for Rhea and the upturn in the curve for Tethys are due to Population I, whereas the peaks at about 10 km diameter for Mimas, Dione, and Tethys are due to Population II. Some but not all experts recognize a similar pair of populations on the moons of Uranus, the curve for Oberon being essentially a Population I trace.

A reasonable explanation for the Population I craters is that they were formed by sweeping up postaccretional debris, possibly ending around four billion years ago, and are more or less equivalent to the late heavy bombardment documented in the inner solar system. Their size–frequency distribution is not entirely dissimilar to that of the lunar highlands, the degree of difference being quite reasonable in view of their vast separation in space, one result of which would have been to give a preponderance of icy rather than rocky material in the impacting population. Population II craters represent a discrete episode of cratering, with a size–frequency distribution that cannot be reconciled with anything seen at Jupiter.

They are usually attributed to collisions with a swarm of debris in orbit about the planet. Such a debris swarm could have been created by a single major impact within the satellite system, which completely destroyed a satellite or at least knocked large chunks out of one. It follows that there is no reason why the Population II impactors at Saturn and Uranus should have been around at the same time as each other.

If this explanation for Populations I and II is correct, then the cratering by cometary impact that must continue to the present day (sometimes known as Population III), and that must have carried on while the Population II flurry or flurries of activity occurred, is more closely related to Population I than II. Unfortunately, the size–frequency distribution at Jupiter (as expressed by Callisto and Ganymede) is at its most distinctive for crater diameters greater than those possible on the moons of Saturn and Uranus (which are much smaller worlds), so it is difficult to tell how the impactors that caused them compare with Population I. It is possible that the crater population observed on Callisto obliterates the traces of an original epoch of Population I cratering.

In summary, relative dating between one satellite system and another, and between the inner and outer solar system, is fraught with caveats and uncertainties. The sanguine view is that the problem may never be solved, even when the badly needed higher resolution images are available to supply the data that are often lacking in the 1–20 km range, but at least craters give us some idea of the external influences that have molded the shapes of many of the worlds in the solar system. However, having covered the generalities, we are at last in a position to examine individual worlds in detail. Callisto, the largest icy moon in the solar system to fall into the heavily cratered, "dead worlds" class will be our starting point.

5.3 CALLISTO

Callisto is the second largest and least dense of the galilean satellites (table 1.1). Gravity measurements by *Galileo* indicate that Callisto's interior is a weakly differentiated ice–rock mixture, more consistent with the right-hand alternative in figure 2.5 than with the left-hand option. There could be a rock-rich core occupying up to about 25 percent of Callisto's radius and surrounded by a mixed ice and rock mantle, but a metallic core would seem to be ruled out. This is perplexing because *Galileo* also found that Callisto has a magnetic field with variable orientation that appears to be induced by interaction with Jupiter's powerful magnetic field. The seat of this magnetic field within Callisto may be a liquid brine layer (a kind of subsurface global ocean), but this is likely to be at a depth of well over 100 km in view of the apparent lack of signs of internal activity affecting the surface.

Galileo and *Voyager* images reveal Callisto as an intensely cratered world (fig. 5.9), although with a dearth of craters in excess of about 60 km diameter and surprisingly few craters smaller than about 1 km across. Callisto's surface is one of the darkest known among the icy satellites, with an albedo (a measure of reflectivity) of about 0.2. If we accept the great age of Callisto's surface implied by its crater density, then some of the darkening can be explained by the presence of a residual "lag deposit" rich in rock dust left on the surface of the regolith as ice

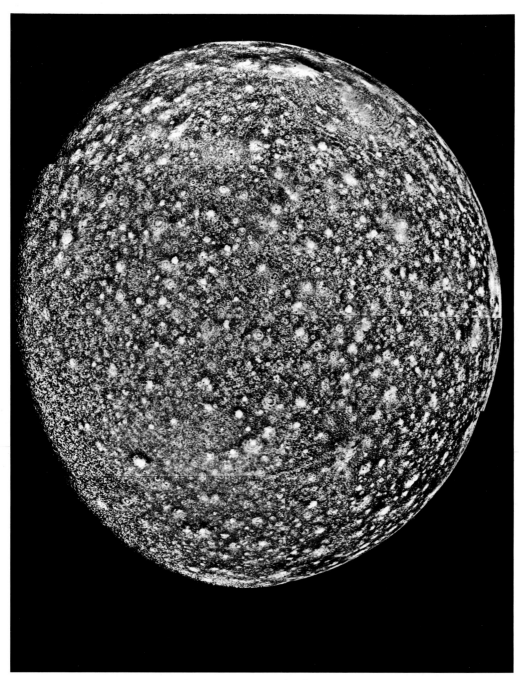

Figure 5.9. Mosaic of *Voyager-2* images of Callisto.

Figure 5.10. Craters on Callisto. The youngest craters are bright, and progressively older craters fade into the background. The largest craters here (which have central pits) are about 60 km in diameter.

has selectively been removed by vaporization during crater-forming impacts, by sputtering, and by sublimation. Reactions caused by solar radiation and charged particles are presumably at least partly responsible, because in addition to the ozone trapped within the ice and forming an insubstantial atmosphere, *Galileo* detected the outflow of escaping hydrogen atoms expected as a result of the breakdown of water-ice molecules.

The younger craters and the rays and other ejecta emanating from them have higher albedos than the average, around 0.4, that fade into the background with increasing age (fig. 5.10). This indicates the penetration of impacts into a more ice-rich (but still impure) subsurface and dispersal of this fresher material in the ejecta. However, even the youngest ejecta material is darker than on other icy satellites, suggesting the presence of a significant proportion of rocky material near the surface. This provides further evidence against differentiation within Callisto, at least in its outer part.

5.3.1 Multiringed Basins

In addition to its craters, Callisto has several multiringed structures, the four largest of which are located in its leading hemisphere, which is more likely to suffer impacts by debris not in orbit around Jupiter. The second largest of these, named Asgard (in keeping with the far-northern name convention chosen for Callisto), shows up rather poorly in the top right of figure 5.9. Figure 5.11 shows the largest structure, Valhalla, somewhat more clearly. The outermost ring of Valhalla is some 2000 km from its center, giving it a diameter of about 4000 km.

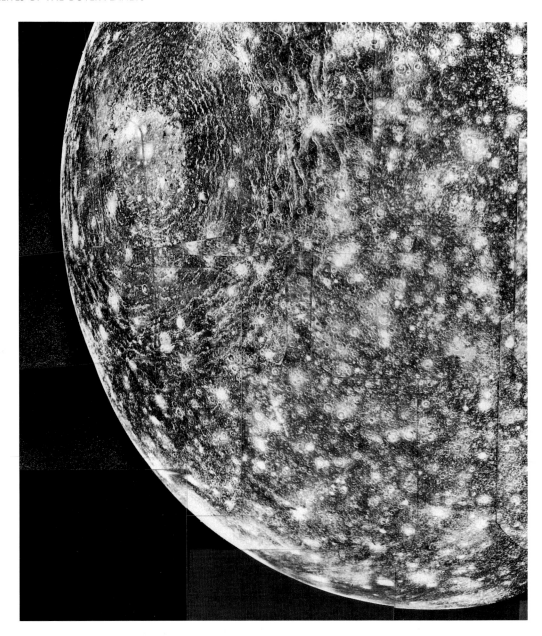

Figure 5.11. Mosaic of *Voyager-1* images, showing most of the 4000 km diameter Valhalla basin on Callisto. Compare with the airbrush map of this feature in figure 5.7c.

This makes it the largest multiringed basin in the solar system; in comparison, the Procellarum, South Pole–Aitken, Imbrium, and Orientale (fig. 5.12) basins on the Moon are 3200, 2500, 1500, and 930 km in diameter, respectively, whereas the Caloris basin on Mercury reaches only 1300 km in diameter.

Despite the similarity of the multiringed basins on Callisto to those on the Moon and Mercury, there are many subtle differences that serve as clues to the strength and thickness of Callisto's lithosphere at the time of their formation. In the center of each of Valhalla and Asgard there is a bright region, 600 and 230

Figure 5.12. Mosaic of *Lunar Orbiter* images, showing the 930 km diameter Orientale basin on the Moon. Compare with the Valhalla basin on Callisto in figure 5.11.

km in diameter, respectively, that is taken to be the remains of a central structure that collapsed under its own weight, either because it was warm and mushy when it formed or because the lithosphere was too thin to support its weight. The central part of the pale area appears flat at *Voyager* resolution, and to describe such a feature as this, which has lost its original morphology and is identified essentially by its anomalous brightness and smoothness, planetary scientists have appropriated the term "palimpsest," which originally meant an ancient parchment on which the original text had been erased and overwritten. The palimpsest at the center of Valhalla is particularly well seen in figure 5.11.

The relative ages of the multiringed basins are given by their relationships to smaller craters; for both Asgard and Valhalla the central palimpsests have about one-third of the density of superimposed craters visible on the surrounding terrain, and of those craters that intersect a ring about one-third are clearly younger than the ring, whereas the remainder are overprinted by it. *Galileo* images at 60 m resolution show much fine-scale detail within the Valhalla palimpsest but fewer really small impact craters than elsewhere on Callisto, suggesting that whatever process acts to remove these has been especially efficient here.

Although the palimpsests in the centers of Callisto's basins contain no direct evidence of their origin, plenty can be inferred from the concentric rings around each basin. In the case of Valhalla the rings are about 15 km wide and are 20–30 km apart in the inner part of the basin, rising to 50–100 km apart in the outer region. No single ring runs completely around the structure; instead, they have the shape of sinuous arcs that are mostly between 200 and 500 km in length. For a distance of 200–300 km outward from the bright central region the rings consist of ridges less than 1 km in height that are paler than the terrain in which they lie. Beyond this the rings are dominantly outward-facing scarp slopes 1–2 km high, although a few are inward-facing scarps and some near the inner edge of this

Figure 5.13. Mosaic of two *Galileo* images showing an area 38 km across on Callisto, illumination from the left. The scarp of one of the Valhalla rings runs obliquely across the right-hand portion of this view, and younger impact craters down to less than 200 m across can be seen, of which there are surpisingly few. Note that only the summits of ridges appear bright and are presumably clean ice, whereas most of the area gives the appearance of being mantled by a dark dusty lag deposit. This is typical of areas of Callisto imaged in high resolution by *Galileo.*

outer zone even take the form of troughs. Where a ridge cuts a preexisting crater, the crater remains circular with no measurable shortening or stretching. Similarly, when a trough cuts through a crater, the portion of the crater within the trough appears to have been simply down-dropped, with no change in its overall shape, in a structure that a geologist would call a graben. A pale hummocky unit hugs the base of many of the scarps and probably represents material extruded as a melt or as a crystal-rich icy mush along these fractures. The same unit appears to have inundated the down-dropped portions of some of the craters within the troughs. Several of the outward-facing scarps in the Valhalla ring system are visible running from the top to the bottom of figure 5.10, and a detailed *Galileo* view is shown in figure 5.13.

There have been several attempts to explain the morphology of the Valhalla basin in terms of the response of Callisto's lithosphere and asthenosphere to the formation of a large transient cavity caused by a major impact. Essentially, the thinner the lithosphere, the more rings are expected and the closer their spacing. The rarity of radial features in Valhalla provides additional clues to strengths and thicknesses. The details need not bother us here, but it appears that Valhalla is best explained by an impact into a lithosphere that was only about 30 km thick, the strength of which was intermediate between that of the Ross Ice Shelf (Antarctica) and that of Earth's lithosphere. A liquid asthenosphere can be ruled out because that would have resulted in a very different morphology, and it seems clear that the asthenosphere deformed by solid-state flow. A schematic model is shown in figure 5.14; in response to the formation of a transient cavity, the nearby asthenosphere flowed inward to fill the void. This exerted an inward drag on the base of the lithosphere, causing it to fragment into concentric zones. At the same time, subsidence of the raised rim drove a weak asthenospheric flow away from the crater. The inward-flowing region gave rise to the inward-facing scarps and occasional grabens, and the ridges closer to the center are interpreted as extrusions derived through concentric fractures in the lithosphere. The surrounding, weaker outward flow produced the outward-facing scarps.

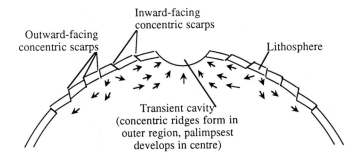

Figure 5.14. Model for the generation of multiringed basins on Callisto, showing the asthenospheric flow induced in response to the excavation of a transient cavity by a major impact, and the concentric fractures that this flow causes in the overlying lithosphere.

One complication to the generalizations about the thickness of an icy lithosphere that were made in chapter 4 is that the deformation behavior of ice depends not only on the imposed rate of deformation (the *strain*) but also on the magnitude of the deforming force, that is, on the *stress* (in particular, the shear stress). The stresses caused by the presence of a transient cavity are so high that the lower part of what would otherwise be called the lithosphere behaves as part of the asthenosphere. The 30 km thickness quoted for Callisto's lithosphere at the time of the formation of Valhalla is thus applicable to the impact-defined lithosphere only. Internally generated (geological) stresses would almost certainly be less, and the effective lithosphere for all but the virtually instantaneous excavation of deep cavities by impacts would be greater than this by a factor of at least two or three.

5.3.2 Crater Relaxation and Retention

Callisto contains abundant evidence of the cooling and thickening of its lithosphere (however defined) through time. The topography of most craters in all size ranges is subdued, and the effect is greatest for the oldest and largest craters. This is presumed to have occurred by a process known as viscous relaxation in which the lithospheric ice deformed very slowly under its own weight, tending to even out the variations in topography (fig. 5.15). To do this it must have been warmer than it is today, and perhaps thinner as well. The time required to virtually wipe out the topography of a crater 20 km in diameter could have been of the order of a billion years, but relaxation could have been much quicker immediately after impact if local impact heating was adequate. The viscous relaxation of Callisto's early large craters so that they disappeared completely is probably a contributing factor to the deficiency of larger craters in its crater size–frequency distribution.

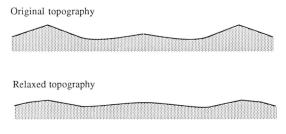

Figure 5.15. Relaxation of crater topography, such as occurs when a relatively large crater is formed in a warm, thin lithosphere. This effect can be seen in many of Callisto's craters.

The relaxation of craters takes on a special significance when trying to interpret Callisto's impact history and lithospheric evolution. The four largest multiringed structures are in the leading hemisphere, and craters more than 60 km in diameter are twice as abundant near its apex of orbital motion (i.e., in the middle of its leading hemisphere) than near the antapex of motion, but there is no leading–trailing asymmetry in the density of 1–10 km diameter craters. If the impacting bodies originated externally to the Jupiter system, nearly ten times as many craters would be expected at the apex of orbital motion as at the antapex. One way to explain this is that most of the impactors (especially those forming 1–10 km craters) originated *within* the Jupiter system. An alternative model suggests that there was a leading–trailing asymmetry in cratering but that most of the evidence has been lost as a result of asymmetric strengthening of Callisto's lithosphere over time. In a strengthening lithosphere the oldest craters to survive should be the smaller ones (1–10 km diameter), and the proponents of the asymmetric cratering model for Callisto claim that the oldest survivors of this size occur around the antapex of its orbital motion. This implies that the lithosphere cooled and thickened faster there than in the leading hemisphere, where the record of such ancient craters has been lost due to viscous relaxation.

Slower cooling and thickening in Callisto's leading hemisphere can be explained if it was blanketed by a thicker layer of regolith, the thermally insulating properties of which would keep the local lithosphere warmer and thinner. A thicker regolith on the leading hemisphere is exactly what would be expected as the result of a greater number of impacts in that hemisphere by impactors derived from beyond the Jupiter system. Similar asymmetries in the supposed crater retention ages and crater densities on Ganymede support this view. If true, this model implies that Callisto was in synchronous rotation from a very early time (which would not be surprising in view of the tidal despinning model discussed in chapter 2) and that the lithosphere was fixed in place with respect to the globe as a whole, unable to slide around freely on top of the asthenosphere.

Despite the indications of a surviving subsurface global ocean suggested by Callisto's magnetic field, its lithosphere appears by now to have grown too thick to allow the formation of multiringed basins even in response to the biggest impacts; this is shown by the presence of one large, bright, fresh crater with a diameter of about 150 km and with no sign of rings around it like those that were produced by inward collapse into Valhalla, Asgard, and the other older basins.

Setting aside the unresolved arguments about the source of the impacting bodies and the degree of leading–trailing asymmetry in cratering, we can summarize the evolution of Callisto, as indicated by the nature of its cratering record, as follows. After accretion and apparently very little differentiation, the lithosphere was at first too warm to retain any recognizable craters. This elevated temperature was due to a combination of residual heat from accretion and heat generated by tidal despinning and radioactivity. With time, as the heat from the first two sources leaked away and the rate of radiogenic heat production declined, the lithosphere cooled and thickened sufficiently to allow the retention of small and then progressively larger craters. As we will see in chapter 6, this is a story repeated at Ganymede, but with the addition that Ganymede is differentiated and has experienced major tectonic resurfacing.

Plate 1. Composite of *Voyager-2* images showing a close-up view of part of Miranda with a resolution of less than 1km, silhouetted against the planet Uranus. Arden Corona lies in the left-hand region of Miranda, with cratered terrain to the right. An interpretation of the faulted structure of the edge of the corona is given in figure 6.37.

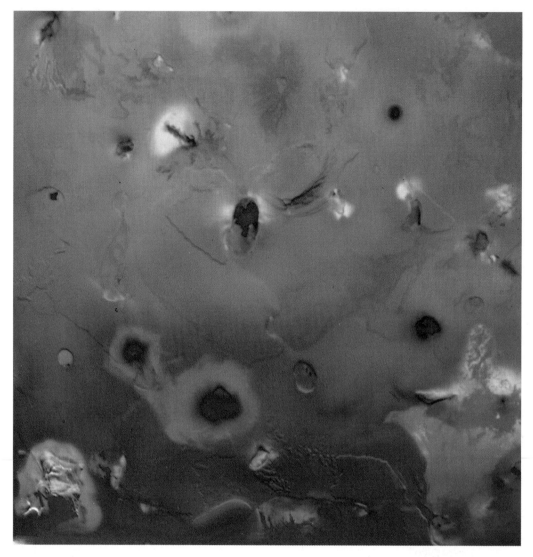

Plate 2. *Voyager-1* image of a characteristic part of Io's surface, in high southern latitudes, 1800 km across. Layered plains material occupies much of the image. There are many vents, notably the shield volcano Masaaw Paterea (*top center*) and below it the elongated floor of Creidne Patera partly covered by dark flows. Immediately east of this, the northwest-ward-tilted mountain block of Euboea Montes rises about 10 km above the plains. The prominent mountain in the lower left corner is Haemus Mons.

Plate 3. *Voyager-1* view looking toward the limb of Io, showing Pele in eruption. The vent is in the rugged volcanic complex toward the upper left. The plume is visible only against the background of space, where it has been enhanced to make it show up.

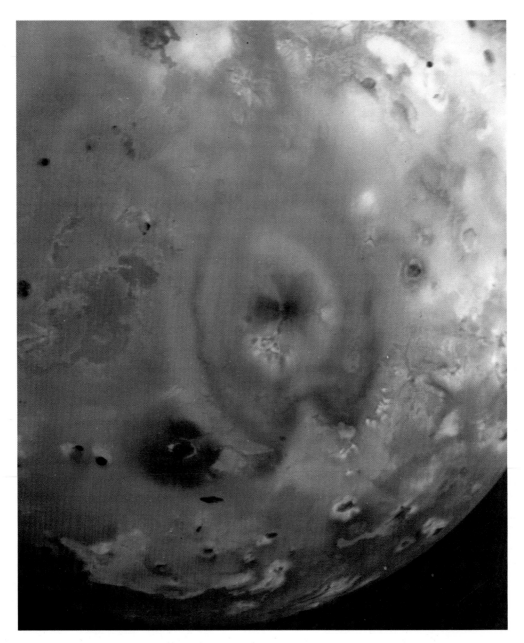

Plate 4 *Voyager-1* image (about 1850 km across) looking down on the surface of Io through the Pele plume. The plume itself is virtually transparent, but the dark concentric deposits resulting from it can be clearly seen. By the time *Voyager-2* was able to image this area, the "bite" missing from the southern rim of the deposit had been filled in, but the plume was no longer active. However, Pele remained a prominent infrared hot spot on Io during the succeeding decades, and plumes were detected there using the Hubble Space Telescope in 1995 and 1996 that remained active during the *Galileo* encounters.

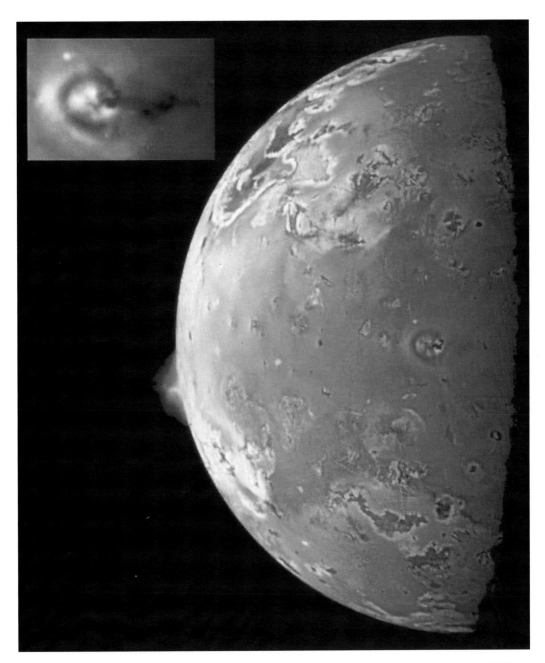

Plate 5. *Galileo* image of Io recorded in June 1997. There are two eruption plumes visible. One is 140 km high and seen in profile above the limb; this emanates from the caldera of Pillan Patera. The other plume, from Prometheus, is seen from directly above, and lies near the center of the disk. It is shown enlarged in the inset at the upper left; the bluish dark ring is the outline of the plume, and this casts a reddish shadow over the surface to its right. The dark flowlike feature extending to the right of the plume source was not present in 1979.

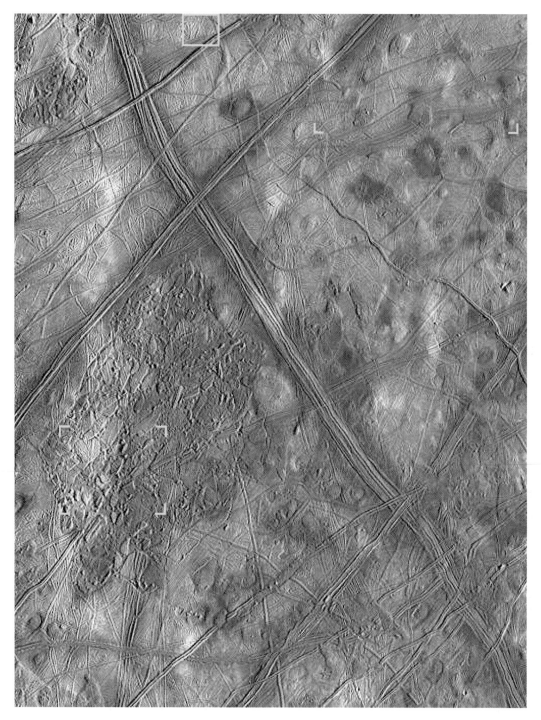

Plate 6. Enhanced color image of Europa (200 km across and centered at 10° north, 270° longitude), made by combining near-infrared, green, and violet *Galileo* images in red, green, and blue, respectively. In this rendering, blue is the general color of the icy surface and red areas are ice-poor (probably salty) regions. The white patches are covered by fine ejecta from an impact crater 1000 km to the south. The prominent ridges intersecting above left of center were mapped as triple bands by *Voyager*. The yellow box locates figure 7.27, and the other yellow marks indicate the lower corners of figure 7.30 and all four corners of figure 7.31.

Plate 7. Mosaic of the highest resolution images of Triton. The equator runs more or less from top to bottom through the center of the plate. The south polar cap of nitrogen-ice occupies the left side, cantaloupe terrain occupies the upper right, and below this are "smooth" high plains (near the terminator) and hummocky terrain (between the smooth plains and the polar cap). A terrain map of the area covered by this mosaic is given in figure 7.39.

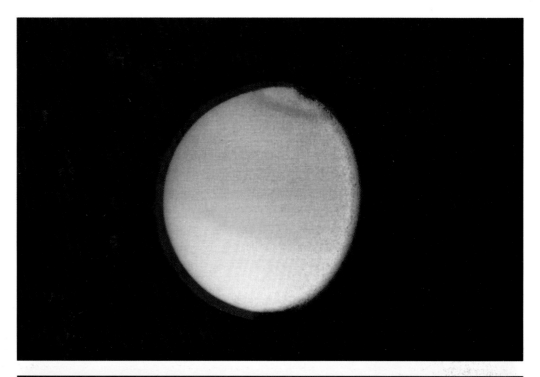

Plate 8. Two *Voyager-1* views of Titan. The red color is due to traces of nitrogen-bearing organic molecules within the opaque aerosol layer, 200 km above the ground surface. (*Top*) General view showing the contrast in brightness of the atmosphere between the southern and northern hemispheres and the dark north polar hood. (*Bottom*) Blue haze layers seen above the limb.

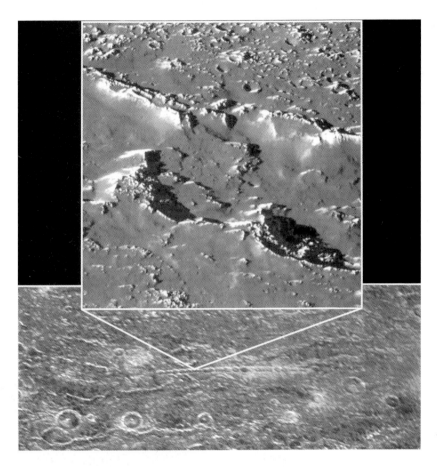

Figure 5.16. A chain of twenty-five 10 km diameter craters on Callisto believed to result from the serial impacts by fragments of a split comet. (*Top*) Detailed oblique view from *Galileo,* covering an area only 13 km across and showing parts of three overlapping craters in the chain. As in figure 5.13, bright, clean ice is exposed only on topographic eminences. (*Bottom*) *Voyager* image showing the whole chain in context.

5.3.3 Crater Chains

There are a dozen or more linear chains of impact craters on Callisto, such as the one shown in figure 5.16, whose origin was not understood until the discovery of comet Shoemaker-Levy 9 in 1993. In the previous year this comet had passed, unobserved, within 90,000 km of Jupiter (barely one planetary radius away), where tidal forces had ripped it into about twenty sizable fragments. The comet's trajectory then took these fragments 55 million km away from Jupiter before returning to crash serially and spectacularly onto Jupiter during a week-long period in July 1994. Comets probably pass close enough to Jupiter to be tidally fragmented, on average, about once every hundred years. However, occasionally one of Jupiter's satellites must happen to lie in the outbound path of the newly disrupted fragments, in which case the rapid succession of impacts onto the satellite will produce a crater chain. Virtually all the crater chains on Callisto lie, as expected, in the Jupiter-facing hemisphere. They are up to a few hundred kilometers long and are composed of craters ranging from less than ten to about forty kilometers across.

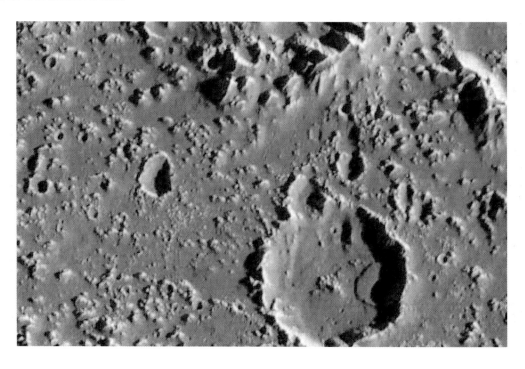

Figure 5.17. *Galileo* image showing a 44 km wide area of Callisto including a 10 km diameter crater at the lower right, part of whose eastern wall has shed a landslide across the crater floor.

5.3.4 Landslides and Frost Migration

Callisto today is essentially an inert world, because of the coldness and thickness of its lithosphere. However, this does not mean that the next time a 10 km diameter crater forms on Callisto it will necessarily just sit there, without relaxing, indefinitely, until obliterated by subsequent craters. The high resolution provided by *Galileo* images revealed a variety of degradational crater forms, many of them attributable to landslides (fig. 5.17). The failure of the crater walls implied by the landslides could be a result of their structure having become weakened because ice has been lost by sublimation, impact vaporization, and radiation breakdown (processes which, as we have seen, liberate oxygen and hydrogen), so that the resulting dust-rich material cascades downslope. Alternatively, the landslides could be triggered by nearby impacts.

Callisto's landslides must be very infrequent events. So what would a visitor find happening on Callisto? There are towns on Earth where they say the most exciting thing to do on a Saturday night is to go out and watch the traffic lights change. The equivalent on Callisto would be to go and watch the frost migrating. Figure 5.18 shows a portion of Callisto's surface at high northern latitudes. The inside walls of most of the craters appear brightest on the south side of the crater. This is clearly not an illumination effect because these slopes face away from the Sun, and it is the opposite slopes, on the south-facing inside wall on the north side of each crater, that are the most fully illuminated. *Galileo* revealed a similar effect in high-latitude craters on Ganymede.

The midday temperature on Callisto is about 116 K at 60° N. This is sufficient

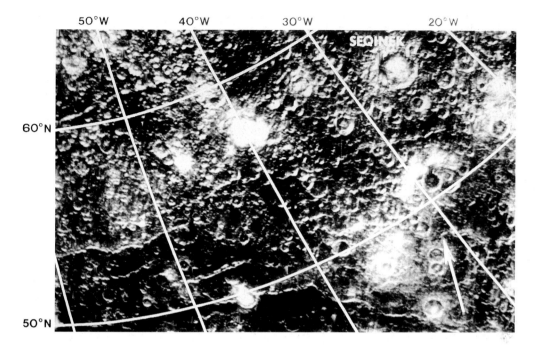

to allow a significant amount of sublimation of ice (i.e., molecules escaping directly from the solid to the vapor phase) over geological time. It has been calculated that a south-facing slope of a crater at 60° N would lose ice by sublimation at a rate of about 60 m per billion years, until the process became checked by the concentration of a dusty residue (a "lag" deposit) over the surface. In contrast, sublimation from a north-facing slope, where the temperature is probably some 30 K lower, would be at less than a thousandth of this rate. A possible likely explanation of the whitening of north-facing crater walls is that ice has sublimed in the more fully sunlit and therefore warmer areas nearby and has been preferentially redeposited on the colder, north-facing slopes.

A similar explanation can be offered as to why only the ridges and peaks are bright in most areas of Callisto that have been imaged at high resolution (e.g., figs. 5.13 and 5.16). Peaks are more exposed than valleys to the sky, so peaks tend to be slightly colder. This means that sublimation of ice is greater in valleys and condensation of frosts is greater on peaks. The relative differences are slight, but once the process has started there is a feedback loop that increases its efficiency. Areas of frost condensation will have higher albedo, which means they will reflect more of the incident solar radiation and so remain cold and be effective frost traps. In contrast, areas where ice sublimes will, by virtue of the lag deposit of dust left behind by the vanished ice, have lower albedo and so warm up more in the sunlight, which will accelerate the rate of ice sublimation. This is an attractive theory, but sublimation of ice would become impossible by the time the surface dust layer became more than a few particles thick, so it fails to account for the apparent thickness of the dust-rich material on the images. Perhaps stirring up of the top few tens of meters by small impacts ("impact gardening") enables, over time, all the ice originally above that depth to be brought within the

Figure 5.18. A high-latitude region of Callisto showing the frost-induced brightening of the north-facing crater walls. This is particularly well shown within the 60 km diameter crater Seqinek, toward the upper right. The illumination direction is shown by an arrow.

reach of sublimation. However, the paucity of small craters on high-resolution images still defies explanation.

We now turn to the other heavily cratered "dead worlds" out at Saturn and Uranus, where the surface temperatures are probably too low to allow even the limited relief to monotony provided by sublimation and frost migration. As we have already seen, their differing crater size–frequency distributions indicate at least two populations of impactors. Aside from that, the main interest in these other heavily cratered worlds lies in how and why they differ from their more recently active neighbors.

5.4 RHEA

Rhea lies in the middle of Saturn's family of moons (fig. 1.2). With a radius of 764 km, it is considerably smaller than Callisto but is the largest of the other dead worlds. The bulk density measurements made during the *Voyager* flybys suggest that the composition of Rhea is about 40 percent rock if it is a differentiated world, or about 35 percent rock if it is undifferentiated, the remainder being water-ice with possibly a small fraction of methane and ammonia. Its albedo is high, about 0.6, suggesting that it has a much cleaner icy surface than Callisto. Most of its Saturn-facing hemisphere was imaged at a resolution of 20 km, increasing to better than 2 km in the north, but the anti-Saturn hemisphere was much more poorly covered (resolution about 50 km), so we know much less about it. In the well-imaged areas, Rhea appears very much the archetypal heavily cratered icy satellite. In general, large craters of all ages show no

Figure 5.19. *Voyager-1* image of part of Rhea's north polar region, about 300 km across. The craters show no evidence of viscous relaxation of their topography. The polygonal outline of some of the craters may be due to impact into a deep rubble layer. Very bright slopes on some inner walls of the craters may be relatively clean ice exposed by slumping.

evidence of viscous relaxation of their morphology (fig. 5.19; see also fig. 1.4), offering evidence of more rapid cooling and thickening of the lithosphere and also of Rhea's lower gravity compared to Callisto.

As shown in figure 5.8, the most densely cratered region of Rhea is dominated by Population I craters, with a positive slope on the relative size–frequency distribution plot. However, there are regions such as near Rhea's north pole where Population II craters predominate, suggesting a resurfacing event toward the end of the Population I bombardment that some researchers have attributed to mantling by ejecta from two poorly imaged large (possibly multiringed) basins, and others to episodes of some kind of endogenic resurfacing. Possible clues that there was endogenic activity on Rhea are given by a few linear troughs that run through Population I terrain but that cannot be traced into younger Population II areas. Kun Lun Chasma, shown in figure 3.10, is an example. These could record an episode of thermal expansion as Rhea heated up, or a volume change associated with differentiation (although they seem to be too young for both of these), but could equally well be impact-generated fractures.

The general lack of evidence of geological activity on Rhea is surprising in view of its probable thermal history since formation. Two models for Rhea's internal temperature evolution over time are shown in figure 5.20, differing essentially in how much accretional heat is retained. Warming beyond the initial state is modeled on the basis of radiogenic heating, using the ice-to-rock ratio derived from the bulk density measurements of the *Voyager* experiments. On both models the interior warmed up (although never to melting point) and began to convect in the solid state about a few tens of millions of years after formation, and continued to convect to within 1.5 billion years of the present day, by which time the whole globe had become effectively lithospheric. The freezing of any traces of liquid ammonia hydrate would have been complete at about the same time. Presumably, subsequent cratering has obliterated traces of surface tectonism (with the dubious exception of the linear troughs) that occurred while the lithosphere was still thin enough to be fractured and deformed by internal convection, but this makes it difficult to explain the clear signs of tectonism on many of Saturn's smaller and generally less dense satellites described in chapter 6.

Another puzzle posed by Rhea is why there are no signs of any global contraction in volume. As noted in chapter 2, Rhea is large enough for ice II to be the stable phase near its center. This is the case whether we accept an undifferentiated (fig. 5.21a) or a differentiated (fig. 5.21b) model for Rhea's internal structure. Referring back to the ice phase diagram in figure 2.4, it can be seen that the ice II–ice I phase transition shifts toward lower pressures as the temperature decreases. This means that as Rhea cooled, the central ice II zone would have grown at the expense of the overlying ice I zone. Ice II is about 20 percent denser than ice I, and the conversion of ice I to ice II as the temperature fell should have resulted in a global contraction of Rhea's radius by about 15 km, leading to a reduction in its surface area of 4 percent. This should have occurred gradually during the later part of Rhea's history, but it is difficult to reconcile the observed linear troughs with compression.

To have a better chance of understanding Rhea, we need higher resolution im-

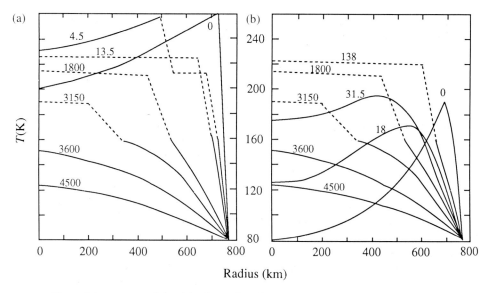

Figure 5.20. Two models of the thermal evolution of Rhea. (*a*) "Hot start," with a large degree of accretional heat retained. (*b*) "Cold start." These plots show the internal temperature profiles at various times, in millions of years after formation. The solid lines indicate heat transported by conduction; the broken lines show heat transferred by solid-state convection. The models are very similar from about 100 million years after formation and show a convective asthenosphere enduring until about 3.3 billion years, below an ever-thickening lithosphere.

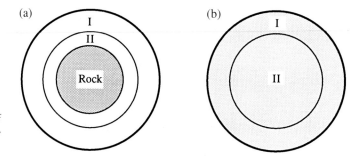

Figure 5.21. Two extreme models for the internal composition of Rhea, showing the stability fields of ice I and ice II. (*a*) Fully differentiated, with a rocky core. (*b*) Undifferentiated.

ages of the anti-Saturn hemisphere, the trailing part of which is darker than the rest of the satellite, but which has pale wisps superimposed on it that could be the missing tectonic features. With the 50 km resolution images available at present, there is little that can be said other than that parts of Rhea evidently underwent some kind of resurfacing event toward the end of Population I time (which may represent the end of the late heavy bombardment), but we do not know what form this resurfacing took. For all we know, in the poorly imaged hemisphere there could be vast tracts of internally reheated and remobilized terrain, or volcanic vents that could have obliterated some of the Population I terrain with a mantle of "pyroclastic" ice. Such features would make it more straightforward to interpret the rather uninformative hemisphere of which we do have good images. For the time being, Rhea must be consigned to the "dead worlds" class, with the remark that in the next chapter we shall see examples of worlds where past episodes of resurfacing are much more obvious.

5.5 IAPETUS

Iapetus, the outermost of Saturn's major moons, is a strange world. In terms of size and density it is a close twin of Rhea (table 1.1) and would be expected to have a similar internal structure and thermal history. Long before the space age, Iapetus was famous for the dramatic difference in albedo between its leading and trailing hemispheres. The leading hemisphere has an overall albedo of only 0.1, whereas that of the trailing hemisphere is around 0.5, a value typical among icy satellites. In consequence, through a telescope Iapetus appears considerably darker when to the east of Saturn in the sky (when its leading hemisphere is seen) than when it is to the west of the planet (when its trailing hemisphere is seen). This is just the sort of anomaly that science fiction writers like to get their teeth into; for example, in the book (although not the film) *2001, A Space Odyssey*, the "Star Gate" monolith erected by alien intelligences was located in the exact center of the bright hemisphere.

The *Voyager* probes did not find anything quite so amazing, although Iapetus remains something of an anomaly and an enigma (fig. 5.22). Unluckily, neither probe was able to pass very close to Iapetus; *Voyager-1* provided images of parts of the Saturn-facing hemisphere with a finest resolution of about 50 km, and *Voyager-2* imaged primarily the anti-Saturn hemisphere rather better with a resolution that reached about 20 km in places. The best images revealed that the bright terrain is heavily cratered for diameters greater than 30 km. Smaller craters were too small to see on the images, but by extrapolating the size–frequency distribution to smaller diameters, the total crater density would seem to be comparable with densely cratered objects such as Mercury, Callisto, and Rhea, and also the lunar highlands. There are no traces of any tectonic activity.

The dark hemisphere is so dark that the *Voyager* imaging system was not able to record any topographic details, such as craters, within it (although this might have been achieved with the kind of long exposures and motion compensation used later at Uranus and Neptune), so the nature and relative age of the dark material remain unclear. What we do know is that the distribution of the dark material is symmetric about the apex of orbital motion, with an albedo of only 0.02 at the apex, decreasing to 0.04 near the edges of the dark terrain. This sort of distribution argues strongly for an external control of the darkening process. Ground-based photometry shows that the dark material is very red, and it has been suggested that its composition can be explained by hydrated silicates with a small admixture of tholins or other organic polymers known to occur in carbonaceous chondrite meteorites. A possible source for this material is impact-eroded dust from Phoebe, a 220 km long irregular satellite in a distant retrograde orbit, nearly four times farther out than Iapetus. Phoebe has a similar albedo to the dark side of Iapetus, but it is less red, so dust from Phoebe may not be the sole component of the dark material on Iapetus. Alternative sources might be other, hitherto undiscovered minor satellites in retrograde orbits beyond Iapetus, or carbonaceous dust from within Iapetus itself that has been preferentially accu-

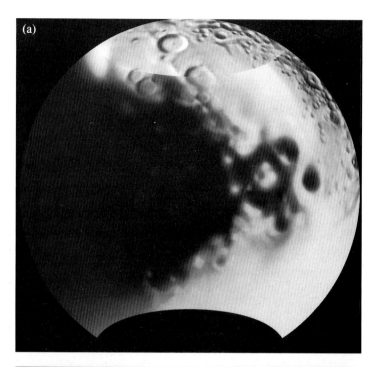

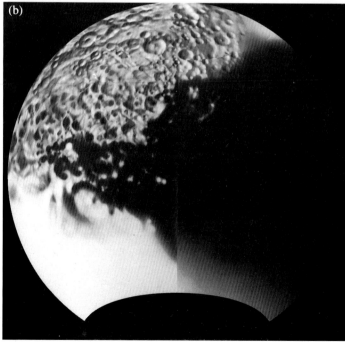

Figure 5.22. Shaded relief maps of Iapetus, showing the Saturn-facing hemisphere (a) and the anti-Saturn hemisphere (b). The leading hemisphere is too dark to allow any details within it to be visible on the *Voyager* images. The unimaged south polar region is omitted. These (and all the other shaded relief maps of complete hemispheres in this book) are equal area projections.

mulated on the surface of its leading hemisphere by impact reworking and volatilization of the surface ice.

If the darkening is due to a surface coating, it must be either very young or very thick because there are no examples of craters penetrating the dark material and depositing paler ejecta on top of it, or indeed of pale spots of any kind within the dark hemisphere. A young age is also indicated by the asymmetry, because Iapetus, being so far from Saturn, is only weakly tidally locked, and large impacting projectiles of the sort that would have formed its largest craters probably carried enough momentum to temporarily change the rate of spin, with the result that the leading hemisphere would have changed its position on each occasion. Over time, this would lead to a roughly even distribution of dust across the surface of Iapetus, so the logical explanation is that the darkening we see is a relatively young phenomenon.

One puzzle that remains with any of the proposed externally controlled origins for the dark material is that, as shown on figure 5.22, there are craters at the edge of the pale hemisphere that have dark floors. Externally derived dark material should coat crater floors, crater walls, and the surrounding terrain equally, and the presence of dark floors only would seem to indicate that at least some of the dark material may come from within Iapetus in the form of cryovolcanic lava extruded onto the floors of these craters. Alternatively, perhaps the dark material has been shed from the crater walls by landslides. Unfortunately, there are no spectral data to confirm whether or not the dark crater floor material has the same composition as the dark hemisphere coating.

It remains a possibility that the dark material has been deposited by volcanic processes. However, if this were the case it would be a remarkable coincidence that the dark material is distributed so symmetrically about the apex of motion. Iapetus then is probably a "dead moon," from a geological if not from a dust-gathering point of view.

5.6 MIMAS

Mimas is the smallest and innermost of Saturn's regular satellites. Being so small, neither accretional energy nor radiometric heating is likely to have been sufficient to drive convection at any time in the past with sufficient vigor to have left traces on its surface, as indicated by the thermal models in figure 5.23.

Mimas has been imaged with a resolution as good as 2 km in much of its southern hemisphere, although the northern hemisphere coverage is markedly poorer. The surface is heavily but not uniformly cratered. Overall, the crater size–frequency distribution for Mimas is dominated by Population II cratering (fig. 5.8), and large tracts of the surface are devoid of craters greater than 30 km in diameter, suggesting that these areas have been resurfaced. Mimas's most striking feature (if you will pardon the pun) is a well-preserved 130 km diameter crater (fig. 5.24), named Herschel after Sir William Herschel, who discovered Mimas in 1789. A few troughs crossing the surface roughly radially to the center of Herschel are evidence of the severe effect of this particular impact, which was probably close to the maximum size that Mimas could have sustained without

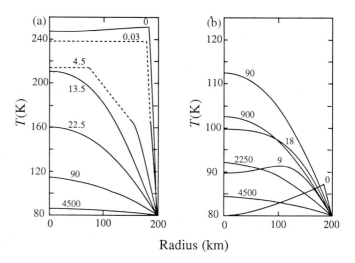

Figure 5.23. Two models of the thermal evolution of Mimas. (*a*) "Hot start." (*b*) "Cold start." These plots show the internal temperature profiles at various times, in millions of years, after formation in the same manner as figure 5.20. Mimas is unlikely ever to have experienced prolonged convection and is almost certainly undifferentiated.

Figure 5.24. *Voyager-2* image of Mimas, showing the enormous crater Herschel, which is one-third the diameter of the moon itself and has a depth of about 10 km and a central peak about 6 km high.

breaking apart. This is a salutary reminder that a satellite can be destroyed if a sufficiently energetic impact occurs. Most of the debris from such an event would subsequently reaccrete to form the satellite anew, but the new surface, and that of its neighbors, would experience a fresh interval of bombardment until the last fragments were swept up. If this has gone on, then we are chasing shadows when we try to construct timescales based on crater statistics in the outer solar system. It is possible that any satellite we see today with a diameter less than about 1000 km has been broken up and reassembled from its own fragments one or more times since its initial formation. Its subsequent differentiation would be much the same as the first time around, except that as time went by there would be pro-

Table 5.1 Two models for the mass fractions of the chemical constituents of the satellites of Uranus. The carbon in model 2 could occur in organic molecules instead of graphite; this would require the rock fraction to be greater, in order to be compatible with the observed bulk density.

Chemical constituent	Model 1	Model 2
Silicate rock	0.336	0.298
Water-ice	0.511	0.455
Ammonia-ice	0.074	0.065
Methane clathrate	0.079	0.070
Graphite	0.0	0.112

gressively less radiogenic heat to assist with the process. As we shall see shortly, impact fragmentation and reaccretion are particularly likely to have happened to the inner satellites of Uranus.

5.7 OBERON

The two remaining "dead worlds" are Oberon and Umbriel, the outermost and third outermost of the satellites of Uranus. The bulk density measurements made of the five major satellites of Uranus during the *Voyager-2* flyby suggested that there are no clear compositional differences among them. However, opinion is divided as to whether the carbon that must be present (unless the present models of solar system chemistry are very badly wrong) is dominantly organic or a mixture of methane and graphite. Some alternative chemical compositions are given in table 5.1.

Despite their apparent uniformity in composition, the satellites of Uranus display a striking range of geological histories, even when comparing moons of a similar size. For example, Oberon has a different history than Titania, which is its twin in terms of size and mass. However, present judgments should be treated with caution, because the orientation of the Uranus system at the time of the *Voyager-2* flyby meant that only the southern hemispheres of the moons were in sunlight, so we have images covering only half of each world. Furthermore, because of the bull's-eye geometry of the encounter (fig. 3.7a), the resolution of the images is poor for the outer satellites.

Voyager-2 revealed Oberon to be a heavily cratered world, within the limits imposed by the 20 km resolution of the best images (fig. 5.25). The surface is generally fairly dark, with an average albedo of about 0.25, but what appear to be the youngest of the larger craters are surrounded by brighter ejecta blankets and ejecta rays (fig. 5.26). This could indicate a dust-rich surface overlying an icier substrate, as suggested earlier to explain a similar phenomenon on Callisto. An alternative explanation is that what was originally methane incorporated in the surface ice (as a clathrate) has been damaged by exposure to cosmic radiation, in particular, high-energy electrons trapped in Uranus's magnetosphere. It has been calculated that bonds between carbon and hydrogen would be broken as a result of this radiation at a rate sufficient to convert any methane in the upper millimeter of Oberon's surface to dark organic compounds (tholins) and carbon in less

Figure 5.25. *Voyager-2* image of Oberon. Note the peak seen on the limb toward the upper left.

than ten million years. This is an attractive idea, in that it explains both the general darkness of all the satellites of Uranus and why methane cannot be detected spectroscopically.

Various other characteristics of Oberon can be seen on figures 5.25 and 5.26. One is the dark infill on some of the crater floors, reminiscent of Iapetus near the edge of its leading hemisphere. The rectangular shape of the darkest patch on the floor of the crater Hamlet (see fig. 5.26) suggests that it may be a dark cryovolcanic (i.e., icy) lava confined within a fault-bounded depression.

Another notable feature is a mountain peak about 11 km high and about 80 km wide at the base, which is seen in profile on the limb, or sunlit edge, in figure 5.25. This is probably the central peak of a viscously relaxed impact structure, and detailed study of the limb profile suggests a subdued outer ring encircling it at about 375 km diameter, which would correspond to the crater rim. If this interpretation is correct, then this is the largest impact crater in the Uranian system. Its relaxed topography puts limits on the strength and thickness of the lithosphere at the time of its formation.

Enhanced images show vague hints of a few gently curved faultlike features running across the disk (represented in the lower part of fig. 5.26), which could be a consequence either of impacts or of an episode of global tectonism. However, it is reasonable to class Oberon as a "dead world," especially in view of its crater statistics.

Figure 5.27 is a relative size–frequency distribution plot for craters on the five major moons of Uranus. Crater counts by different researchers tend to produce different curves, which makes interpreting these data particularly problematic. Taking the data as plotted in figure 5.27 at face value indicates that both Oberon and Umbriel have dense populations of large impact craters comparable with the lunar highlands and heavily cratered terrains elsewhere in the solar system. By analogy with the situation at Saturn, this distribution would be attributed to an

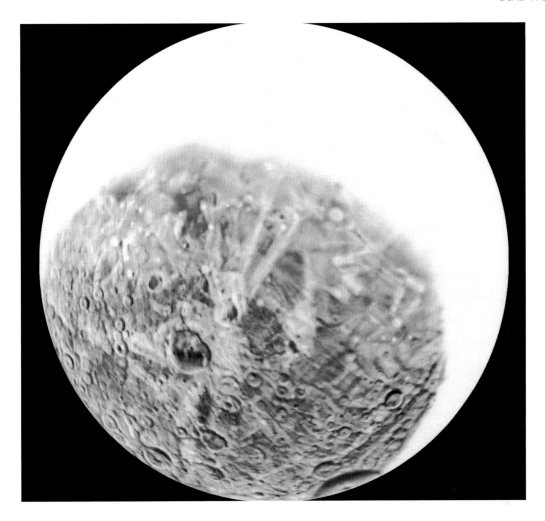

early, Population I bombardment. In contrast, Titania and Ariel have fewer craters, especially at large sizes. This is interpreted by some as representing Population II cratering, and by others as simply younger surfaces recording Population I on which larger craters have been erased, probably by viscous relaxation. The oldest terrain on Miranda is intensely cratered with a peculiar distribution of its own. It is generally believed that the Population I bombardment recorded on Oberon represents the equivalent of late heavy bombardment by postaccretional debris. Calculations show that gravitational focusing of this impacting population would have resulted in a greatly enhanced rate of cratering on the inner satellites, to the extent that Umbriel and Ariel are likely to have been struck at least once during this period (say 4.0–3.5 billion years ago) by a projectile with sufficient energy to have completely disrupted them. The same statistical argument suggests that Miranda would have been disrupted and reaccreted several times, and when Miranda is discussed (in chapter 6), we will see that it may indeed show the aftereffects of such an event.

Figure 5.26. Shaded relief map of Oberon's southern hemisphere, in approximately the same orientation as figure 5.25. Note the bright ejecta associated with dark-floored craters, and the particularly dark patches on the floor of the large (200 km diameter) crater Hamlet, below left of center. The unimaged area has been left blank.

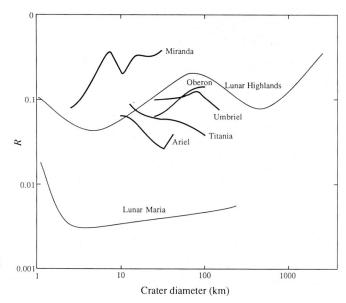

Figure 5.27. Relative size–frequency distribution plot showing the crater populations of the most densely cratered regions on the satellites of Uranus.

5.8 UMBRIEL

Umbriel is smaller than Oberon and, in terms of size and mass, is the twin of Ariel, although their geological histories have been very different. Umbriel's Population I crater record has already been remarked upon. This implies that if Umbriel was disrupted by a major impact during the Population I bombardment, as statistics extrapolated from Oberon suggest is likely, then it must have happened early enough for Umbriel to have reaccreted in time for its new surface to experience a significant duration of continuing Population I bombardment.

As its name implies, Umbriel is a dark world, with an albedo of slightly less than 0.2. The morphology of the craters on its surface shows little or no evidence of any viscous relaxation (Figs. 5.28 and 5.29). Unlike Oberon, or indeed the other satellites of Uranus, Umbriel has no craters with bright rays or other pale ejecta blankets. The only bright material (albedo approximately 0.5) occurs on the flat floor of the 150 km diameter crater Wunda (seen near the left-hand limb in fig. 5.28) and on the central peak of the nearby 110 km crater Vuver. It is possible that the bright floor of Wunda is covered by a cryovolcanic lava in the same manner as some of the dark-floored craters on Oberon. The fact that Oberon "lavas" are dark and Umbriel "lavas" are pale suggests but does not necessarily require a difference in the lava composition; the discrepancy could be explained by differences in exposure age or, more likely, by a difference in the size, shape, or abundance of crystals and/or bubbles in the extruded melt. However, there are no clear examples of eruptions from central peaks documented on any other icy satellite, so it is difficult to conceive of the bright deposit on Vuver's central peak as being cryovolcanic. An alternative explanation is that the central peak uplift has brought to the surface a pale layer that is elsewhere buried to a depth of 10 km or so.

Figure 5.28. *Voyager-2* image of Umbriel, showing its heavily cratered surface. The large crater with a prominent central peak near the upper left is Vuver, 110 km in diameter. The bright-floored crater Wunda is at the extreme left. Umbriel is darker than Oberon (fig. 5.25), but this is not apparent in comparing these images because of subsequent enhancements.

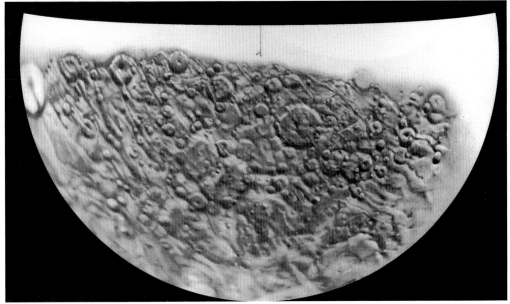

Figure 5.29. Shaded relief map of the southern half of the Uranus-facing hemisphere of Umbriel, covering much the same area as the image in figure 5.28.

Umbriel's overall low albedo is consistent with the radiation darkening of methane with age, although the uniformity of the albedo is surprising. Impacts would be expected to have excavated cleaner, paler material from depth or, if the outer layer is completely uniform to below the depth excavated, at least to have produced a rather paler ejecta simply by fragmentation and pulverization. These problems have led to the suggestion that the generally dark outer layer results from global resurfacing by dark cryovolcanic lavas early in Umbriel's history, predating virtually all the surviving craters. Others have speculated that Umbriel has recently been given a surface dusting by dark material that has hidden whatever surface markings there were previously. This dark dust could have come from a recent impact (a 10 km crater would be sufficient) or an explosive volcanic eruption, although this is not compatible with any observed aspect of Umbriel's surface morphology. Alternatively, perhaps the radiation darkening of methane is entirely responsible for the uniformity of Umbriel's surface, and the persistence of albedo features on the other satellites of Uranus demonstrates the absence of methane in certain regions or at certain depths on those worlds, but its ubiquity on Umbriel.

Inspection of figure 5.29 shows numerous narrow fractures, which cut, and so must postdate, most of the craters that lie along them. Unfortunately, the resolution of the *Voyager* images from which this map was derived, and the limited extent of their coverage, precludes any authoritative interpretation of their origin. However, it is evident that geological processes other than impact cratering must have occurred on Umbriel, even if we don't believe in the cryovolcanic resurfacing of Wunda's floor. With Umbriel we have come to the last of the "dead worlds"; a future space mission may reveal that I have misclassified it and that it really belongs with the recently active worlds described in the next chapter. There we will examine a comparable number of worlds, many of them similar in size, mass, and bulk composition to those in this chapter, but where the record of early cratering has been partly overprinted by the indisputable traces of tectonism and volcanism.

6 Recently Active Worlds

The fashion of this world passeth away
I Corinthians 7:31

With this class of worlds we come to those where a geologist can feel truly at home. Indeed, they excite the curiosity of just about anyone who sees the images, because no longer do we have to be content with counting and measuring craters and pondering the significance of a few dubious fractures. These are the worlds where there is abundant evidence of tectonics on global and local scales, and of vast volcanic outpourings, though even the youngest terrains formed in this way bear the scars of subsequent impact cratering. This means that they postdate the late heavy bombardment, or its outer solar system equivalent(s), but are nonetheless very ancient. If they occurred on Earth these later terrains would be called neither young nor recently active, because they predate the oldest surviving terrestrial landscapes. However, on worlds such as these, without air and running water or younger episodes of tectonics and volcanism, most of their original characteristics have been preserved. We begin with Ganymede (Jupiter), which is the most intensively studied and best imaged example of its class. The other satellites included in this chapter are Dione and Tethys (Saturn), and Ariel, Titania, and Miranda (Uranus).

6.1 GANYMEDE

Ganymede is the largest and most massive of all the planetary satellites. It is bigger than Mercury, although less than half its mass, and is more than twice the diameter and ten times the mass of Pluto (table 1.1). Were it not in orbit around Jupiter, there is no doubt that Ganymede would be regarded as a respectable planet in its own right. Actually, it is not very much bigger than Callisto, and the question of why Ganymede is a fully differentiated body with abundant evidence of tectonics whereas Callisto shows no tectonism and is significantly less differentiated is one of the major unresolved issues in planetary science.

Figure 6.1. Shaded relief airbrush map of Ganymede's leading hemisphere (a) and trailing hemisphere (b), based on *Voyager* images. Changes in the degree of detail shown reflect the variations in resolution coverage (see fig. 3.6). These views show clearly the division into dark and bright terrains, and the yet brighter ejecta from the younger craters.

The wispy markings that can be seen on Ganymede from Earth using large telescopes and that were imaged by *Pioneer-10* (fig. 2.7) were the first hints of Ganymede's strange geology. The higher resolution images obtained during the two *Voyager* encounters showed that the globe is divided into roughly equal proportions between a dark terrain that occurs as polygonal regions and is heavily cratered, and a less heavily cratered bright terrain in the form of bands that separate tracts of dark terrain and in places penetrate the dark terrain in the form of wedges (fig. 6.1). The bright terrain is clearly younger than the dark terrain. It is usually strongly grooved, and provides unequivocal evidence of one or more episodes of widespread-to-global tectonism and volcanism. However, even the bright terrain is fairly heavily cratered, and both it and the dark terrain are peppered with younger craters with bright rims and ejecta blankets. With so many prominent features to name, the stock of available names from the Ganymede myth was quickly depleted, and the majority of names on

(b)

Ganymede have been drawn from the mythology of ancient Near Eastern civilizations.

When *Galileo* got to Jupiter, it sent back high-resolution images of Ganymede showing details smaller than 100 m across within the two main terrain types that caused much *Voyager*-based thinking to be reassessed (fig. 6.2). Moreover, *Galileo* demonstrated that Ganymede has an inner metallic core. Gravity data show that this could be anything from 400 to 1300 km in radius depending on whether it is dominantly iron or iron plus a less dense element such as sulfur. It would occur within the center of the kind of differentiated structure illustrated in figure 2.5. Magnetometer measurements showed that Ganymede generates its own magnetic field. This could conceivably be a result of circulation in an electrically conducting water layer below the ice (as suggested for Callisto), but is probably caused by convective motion in the inner core. This would have to be liquid because solid-state convection is far too slow, and would require a temperature of

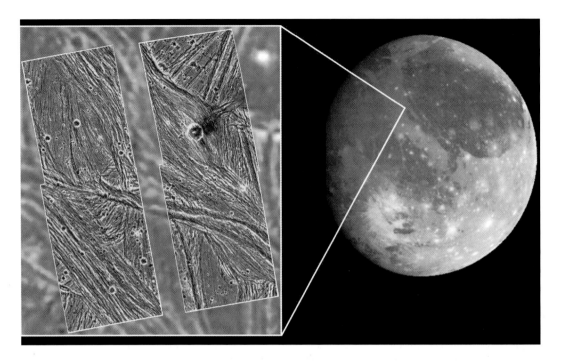

Figure 6.2. A comparison between *Voyager* and *Galileo* images of Ganymede. A 120 by 110 km region imaged by *Voyager* at 1.3 km per pixel has *Galileo* images at 74 m per pixel superimposed. The view on the right (compare fig. 6.1a) is a full disk image by *Voyager* to show the context of the detailed images. The dark terrain in the upper right is Galileo Regio, which can also be seen in the upper left of figure 6.1a.

1980 K at the top of an inner core of pure iron or a more reasonable 1250 K in the case of an iron–sulfur mixture.

6.1.1 The Dark Terrain

We shall consider the dark terrain first, as this has many similarities with, and some notable differences from, the surface of Callisto. In general, Ganymede shows the same sorts of craters as Callisto, except that their topography has on the whole become even more subdued by viscous relaxation, as would be expected on a world that has remained active, and presumably hotter, for longer. The size–frequency distribution of craters on the dark terrain is more or less the same as on Callisto (fig. 6.3). However, whereas craters up to about 100 km in diameter are common on the dark terrain, larger craters are rare. In their place are ancient pale circular patches of subdued relief 100–350 km in diameter, such as Memphis Facula in figure 6.4, that resemble the bright palimpsest in the center of the Valhalla basin on Callisto and are referred to by the same term. Each palimpsest is surrounded by concentrations of smaller craters that are almost certainly secondary craters produced by blocks of ejecta from the main crater-forming incident. Ganymede's palimpsests may be the traces of viscously relaxed major impact structures, as inferred on Callisto. An alternative explanation is that the impacts that produced them broke right through the lithosphere (which would have to have been thinner than about 10 km at the time), and that the palimpsests represent extrusions of warm ice or a slushy ice–water mixture, which was able to escape buoyantly from the asthenosphere and then spread

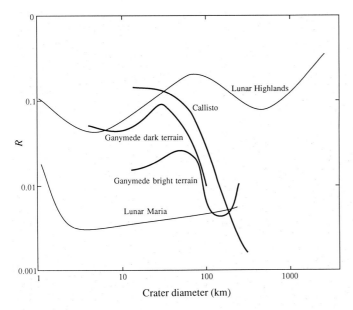

Figure 6.3. Relative size–frequency plot comparing the crater populations on Ganymede and Callisto.

symmetrically across the surface. Figure 6.5 shows a remarkable close-up view of younger impact craters superimposed on Memphis Facula.

Figure 6.3 demonstrates that the crater distribution curve for the dark terrain on Ganymede has much the same shape as the curve for Callisto. This may indicate that, as expected, these two worlds were bombarded by the same population of impactors. An important difference, however, is that Ganymede's curve lies below that for Callisto, showing that the overall crater density on Ganymede's dark terrain is less than the crater density on Callisto by a factor of about three. Other things being equal, Ganymede ought to have more craters than Callisto as a result of the gravitational focusing of incoming projectiles by Jupiter's gravity. The best way to account for the relatively low number of craters on Ganymede's most ancient terrain is that it simply has a younger crater-retention age than Callisto. This could be a result of more effective viscous relaxation on Ganymede or of a global resurfacing (perhaps melting) event. Either of these could be explained by the extra heat generated within Ganymede due to greater radioactive, tidal, or accretional heating, as a consequence of Ganymede being slightly denser and larger than Callisto. Studies of the relaxation of crater topography suggest that Ganymede's lithosphere thickened from about 10 km at the time of the earliest craters preserved (4.0 billion years ago?) to about 35 km by the time the first bright terrain was formed (3.8 billion years ago?).

There are no well-preserved multiringed basins on Ganymede, but there are at least four systems of roughly concentric arcuate furrows, consisting of troughs approximately 10 km wide with raised rims spaced about 50 km apart. These can be traced for distances of several hundred kilometers; part of such a system is well seen in the lower half of figure 6.4. It has been proposed that these systems are of tectonic origin and record an episode of global expansion, but their concentric nature and analogy with Callisto strongly suggest that they were caused

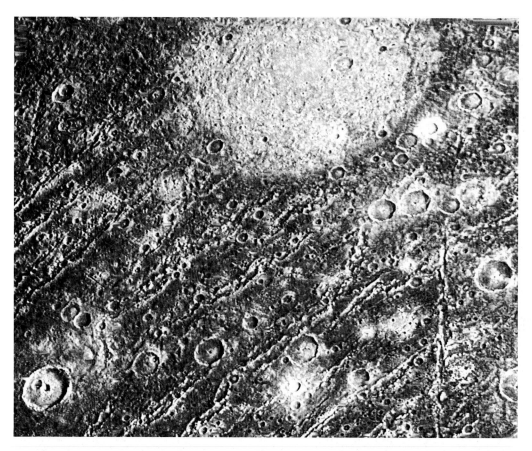

Figure 6.4. *Voyager-2* image covering a dark terrain region on Ganymede, about 700 km across. The 350 km diameter bright palimpsest near the top edge is Memphis Facula. This is either the relic of a large impact crater that formed when the lithosphere was sufficiently warm and thin to allow it to become flattened under its own weight, or a region covered by the extrusion of warm ice when an impact penetrated the still-thin lithosphere. Virtually all the craters in this image are topographically relaxed. An ancient arcuate furrow system trends obliquely across the image. Most of the craters and the palimpsest overprint these furrows, and so the furrows are clearly very old. Memphis Facula is the most obvious palimpsest in the prominent dark terrain of Galileo Regio, and can be made out in figures 6.1a and 6.2.

by impacts. The fact that most large craters overprint any furrow that intersects them shows that the furrow systems originated relatively early in the history of the dark terrain, in which case any central impact structure would be expected to have relaxed to the palimpsest stage or beyond.

Whatever caused them to form, the present morphology of individual furrows can be fairly simply explained, as shown in figure 6.6, by analogy with the topographic adjustments experienced by terrestrial fault-bounded valleys: a trough forms as a graben, by down-dropping of the central block; viscous relaxation in response to isostatic forces then causes the rims and the center of the basin floor to bow upward and the edges of the floor to be depressed. On Ganymede, this

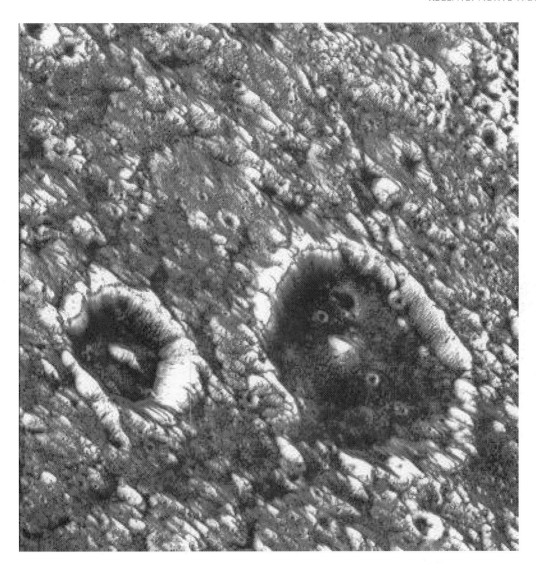

modified topography is likely to persist indefinitely, or at least until obliterated by the superposition of craters, whereas on Earth the raised rims of analogous structures experience rapid erosion, leading to continued isostatic uplift and further erosion of the rims, so the process is liable to continue until the graben has been filled with sediment and the topographic anomaly has been erased.

Many geologists would dearly love to be able to study the effects of plate tectonics on another world. The observation that the dark terrain is broken by bands of bright terrain raises the question of whether the blocks or plates of Ganymede's dark terrain separated by these fractures were ever decoupled from the interior and thus able to slide around in a manner analogous to plate tectonics on Earth. There is a growing body of evidence in favor of lithospheric movements on Ganymede amounting to tens or hundreds of kilometers. These events reached a crescendo with the formation of the bright terrain, but the dark terrain

Figure 6.5. A close-up *Galileo* view of a 23 by 23 km area within Memphis Facula, showing two much younger impact craters. The fine dark line visible near the northern rim on each crater may reveal an uplifted layer from the icy "bedrock." Apart from this, the cleaner surface ice on topographic eminences and concentration of darker (dustier) material in depressions may indicate Callisto-style frost migration.

Figure 6.6. Cross sections to show topographic relaxation of a graben, which may explain the raised
rims on the furrows on Ganymede. (a) Immediately
after formation. (b) After viscous relaxation of topography.

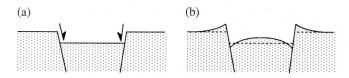

(a)   (b)

did not escape. Figure 6.7 shows a region of dark terrain scarred by several generations of apparently extensional fractures, the oldest being single features that
are cut by younger double or multiple sets, which are in turn cut by a 15 km wide
lane of grooves and ridges. The extensional nature of the latter is demonstrated
by the separation of opposing halves of a 25 km crater in the lower left.

6.1.2 The Bright Terrain

Ganymede's bright terrain is where the geology begins in earnest. It is clearly
younger than the dark terrain, as can be seen by truncation relationships wher-
ever they abut (fig. 6.8) and by the fact that the bright terrain has a lower crater
density (fig. 6.3). In fact, the crater density on the bright terrain varies from place
to place (even after allowing for the possible asymmetry due to Ganymede's orbital motion), which suggests that the bright terrain developed over a period on
the order of half a billion years, probably at some time before 3 billion years ago.

Most of the bright terrain is occupied by belts of subparallel grooves, and indeed, the bright terrain is alternatively known as the grooved terrain. In some
places parts of the bright terrain appear smooth (fig. 6.9), but in many cases
when seen at best *Galileo* resolution even these areas are revealed to be crossed
by ridges and grooves on a scales ranging down to less than a kilometer. Figure
6.10 shows a detailed view of part of the area seen in figure 6.9.

The age range of the bright terrain as determined by crater statistics is compatible with the complex emplacement history of segments of grooved terrain.
The youngest grooved terrain appears to consist of single large grooves or pairs
of grooves, and there are also some groovelike features within the dark terrain
that have no distinct albedo contrast, some of which may be young but others of
which are very old. The bright terrain, then, appears to be the result of a prolonged phase of global activity that built up to a climax and then waned, with
single grooves representing the earliest and latest activity. The most probable origin for the bright terrain is global expansion produced by internal differentiation.
Had Ganymede accreted homogeneously, ice phase changes could result in up to
a 7 percent increase in surface area if all the rocky fraction were able to segregate
into the outer core. This is an extreme model, and as the bright terrain occupies
something like half of Ganymede's surface, it obviously cannot simply represent
extra surface area generated by expansion. Instead, it must be fresh material extruded onto the surface in response to extensional cracking of the lithosphere,
and possibly originating at the sort of subsurface brine ocean hinted at by Ganymede's and Callisto's present-day magnetic fields. This process would be encouraged by thermal contraction of the lithosphere as it cooled, at the same time as
phase changes and/or internal differentiation caused the volume of the globe to
increase. A phase change that could be responsible is the conversion of ice V to
ice II as the internal thermal gradient decreased (fig. 6.11).

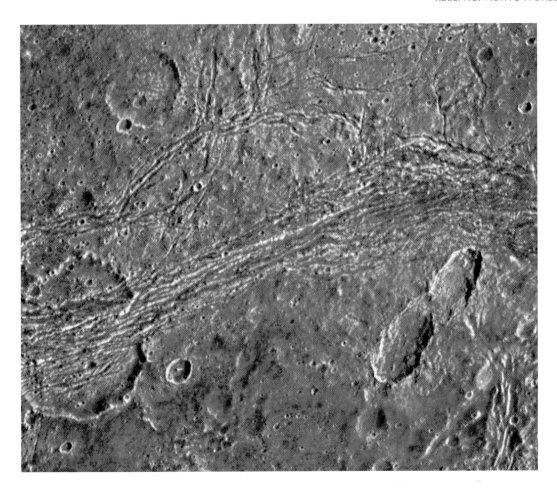

If the bright material was extruded as water, brine, or an icy crystal mush, it would be encouraged to rise through fractures and spill out onto the surface provided that the ice in the lithosphere was made slightly denser than the rising material by the incorporation of rocky fragments (as seems likely). The material extruded from below would be more reflective than the dark terrain because it would contain fewer rock fragments and less dust. There are three different general models that might account for the formation of belts of bright terrain. These, and a variant of the third model, are illustrated in figure 6.12.

In model (a) the belt of bright terrain occupies a gap produced by the tearing apart of blocks of dark terrain lithosphere on either side; the emplacement of the bright material would be analogous to the formation of an ocean basin on Earth by seafloor spreading. There are two main objections to this model that seem to rule it out completely. One is that the implied extension is far greater than can be accounted for, in the absence of any evidence of a nearly equivalent amount of lithospheric compression elsewhere on Ganymede. The second is that although preexisting craters may be truncated by the edge of a belt of bright terrain and although in zones of possible incipient bright terrain formation (e.g., fig 6.7) craters may be split apart, there is no place where the dark terrain on either side of

Figure 6.7. *Galileo* view of a 110 by 90 km area within Nicholson Regio (the dark terrain in the lower left of fig. 6.1b). The two halves of the crater in the lower left have evidently been moved apart by 5–10 km by the processes that generated the lane of ridges and groves running across the center of the image. Older grooves, occurring either singly or in sets, attest to a complex previous history. The pair of elliptical craters near the lower right may have been caused by oblique impact by two fragments of a tidally disrupted comet (compare fig. 5.16).

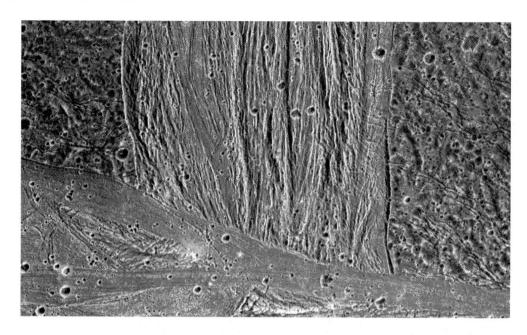

Figure 6.8. *Galileo* view of a 950 by 560 km area of Ganymede showing the relationship between bright and dark terrain. A 500 km wide belt of bright terrain known as Erech Sulcus cuts from north to south through the dark terrain of Marius Regio on either side. Individual grooves within the dark terrain are clearly truncated by Erech Sulcus, which must therefore be a younger feature. However, features within Erech Sulcus are themselves truncated by those in another belt of bright terrain, Sippar Sulcus, that sweeps across the southern part of the image and must be younger still. In detail, both Erech and Sippar Sulci can be seen to be highly complex, indicating that each must have been created by a prolonged series of events. (Erech Sulcus is the narrow belt of bright terrain at the extreme left-hand edge of fig. 6.1a.)

a wide bright belt can be fitted back together again, in a manner analogous to the way in which Africa and South America fit neatly if the South Atlantic Ocean is closed. If you look carefully at the belts of pale terrain converging to a point near the lower left of figure 6.9, you will see that although these are displaced laterally across Byblus Sulcus, they have not actually been moved apart by the width of Byblus Sulcus. Instead, Byblus Sulcus appears to represent the overprinting and obliteration of a zone of dark terrain along the line of a fault with lateral (strike-slip) displacement.

Model (b) shows how a belt of bright material could be supplied by flooding above a crack, which in the case of Byblus Sulcus would be a strike-slip fault. There are sound objections to this model as well; the straightness of the edges of the bright belts and the lack of flooded craters at the edges of these belts strongly favor structural confinement by faults, whereas simple flooding would tend to result in more tortuous boundaries. In addition, the bright terrain material is evidently several kilometers thick in most places, because craters many tens of kilometers in diameter lie within this terrain without excavating dark material from beneath. There are several examples in figure 6.9, but an exception occurs in figure 6.10, where the 8 km crater Nergal has a dark floor and is surrounded by dark ejecta.

The model that best fits the requirement of structural confinement (to account for the straight edges of the belts), thickness of the bright material within the belts, and the small amount of extension possible is that the edges of a bright grooved belt are defined by faults, between which the lithosphere has dropped and across which there may in some cases (as in Byblus Sulcus) have been lateral displacement. The resulting trough could then be flooded by upward-escaping fluids. Model (c) shows this trough as a classic graben, bounded by two faults of equal importance, each of which reaches the base of the lithosphere. The variant

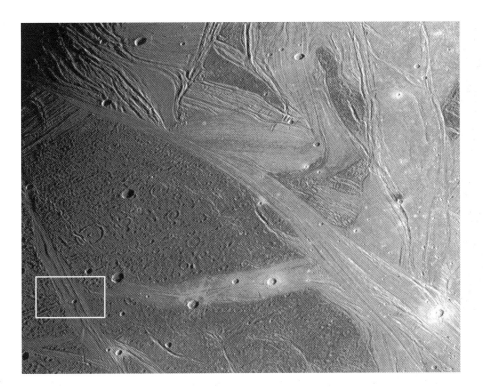

shown in (d) is more in line with discoveries on Earth during the late 1980s that revealed that terrestrial grabens are more commonly formed by extension across a single dominant fault and that the opposite side of such a rift valley is delimited by a less significant fault that joins the major fault at depth.

The nature of the flooding process in the grooved terrain would undoubtedly cast light on the problem of how the bright belts formed, if only we could understand it. Unfortunately, with each increase in image resolution the situation becomes harder to understand. It seems inevitable that the bright terrain was emplaced by some kind of cryovolcanic flooding, but the surface details seem usually to have been modified by local tectonic tilting and viscous relaxation subsequent to flooding. The morphology of the grooves therefore cannot be used to tell us anything about the physical properties of the flow material that was extruded; the grooves and the intervening ridges could be original tectono-volcanic constructs, but they could equally well be the result of subsequent subsidence or rotation of underlying blocks. A few of the possible mechanisms for groove formation are shown in figure 6.13.

Figure 6.14 is a rare example of a *Galileo* image that appears to show evidence of a broad cryovolcanic flow that has not been overprinted by ridge-and-groove formation. The scallop-walled depression could be a collapse feature caused by drainage of a cryovolcanic fluid reservoir, or a result of thermal erosion of the surface during emplacement of the flow that now occupies most of the floor of the depression. Why this area should be so atypical is not clear. The superimposed craters show that it is not particularly young, and the wall of the depression is truncated by a tract of younger pale terrain in the lower left of the figure.

Figure 6.9. *Galileo* image of a 660 by 520 km region containing dark terrain in the north of Marius Regio (top right in fig. 6.1b) and a series of cross-cutting tracts of bright terrain. Various transitional morphologies between dark and bright terrain can be made out, and the bright terrain itself is revealed as highly complex. Note also the subdued relief of most craters in the dark terrain, in contrast to the better preserved shapes of the younger craters in both terrain types. The area in the box is shown in greater detail in figure 6.10, but at this scale it can be seen that the northwest–southeast belt of bright terrain, Byblus Sulcus, that runs through the box appears to displace features on either side by about 60 km in a left-lateral sense.

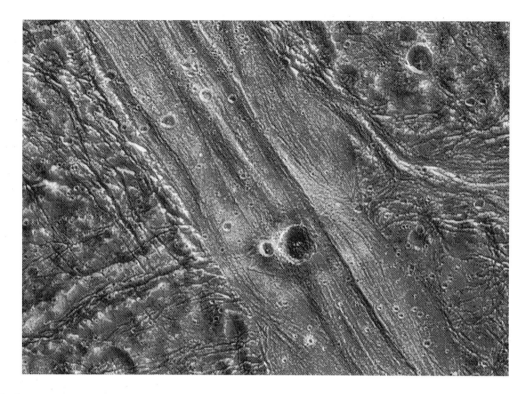

6.1.3 Subsequent History

Figure 6.10. Detailed view of part of Byblus Sulcus, made by combining a high-resolution (86 m per pixel) high-sun *Galileo* image with part of the low-resolution image in figure 6.9 that was obtained with the Sun much lower in the sky and thus revealing the topography better. Complex fractures are revealed in the dark terrain, and the 10 km wavelength ridges within Byblus Sulcus that are apparent in figure 6.9 can be seen to have smaller scale ridges superimposed.

Whatever the mechanism by which the bright terrain developed, the morphology of superposed craters shows that the first ones to form on this terrain became extremely flattened due to viscous relaxation, although later craters have much the same shape as craters of similar age on the dark terrain. This tells a story of cooling and stiffening of the new lithosphere in the bright terrain as it aged.

The thickness of Ganymede's lithosphere today is probably about 300 km, and there is no evidence of changes in global volume since the formation of the bright terrain. This is taken to show that the potential expansion as ice V changed to ice II with cooling was offset by the potential contraction as ice I changed to ice II, in conjunction with general thermal contraction.

The most noteworthy event on Ganymede in post-bright terrain times has been the formation of an impact basin 275 km in diameter that has been named Gilgamesh (fig. 6.15). This is evidently younger than any of the multiringed structures on Callisto, clearly having formed when the lithosphere had already reached a considerable thickness. This makes it a closer analog to the major impact basins on the Moon (e.g., fig. 5.12), Mars, and Mercury, and the blocky mountainous region surrounding the structure is the most rugged terrain known on any of the outer three galilean satellites.

Although Ganymede is the foremost example of a recently active world, as we shall see shortly it is by no means typical. Nowhere in the solar system is there anything that bears close comparison with Ganymede's grooved terrain. Most of the other recently active worlds are characterized by having two or more cratered

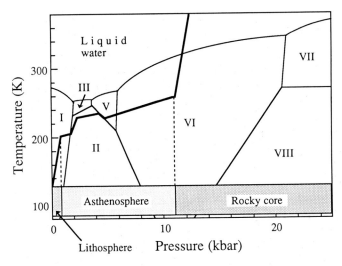

Figure 6.11. Ice phase diagram for a differentiated Ganymede. The heavy line shows the thermal gradient at the time of bright terrain formation. The lithosphere consists of ice I with some rocky contaminants, whereas the convecting asthenosphere contains ices I, II, V, and VI. Parts of the asthenosphere in the ice II region are only about 20 K below the melting point for pure ice and could begin to melt in the presence of dissolved salts or volatiles, thus acting as a source of cryovolcanic magma. As the interior cools, ice V will be converted to ice II, resulting in a slight net expansion. At the present day, the lithosphere is much thicker and extends well into the ice II stability field. The presence of an iron-rich inner core would affect this model very little.

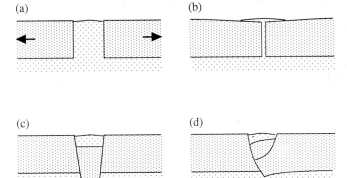

Figure 6.12. Alternative models to account for the formation of Ganymede's belts of bright terrain by the extrusion of water or relatively pure ice slurry from the asthenosphere, in response to extension. See text for discussion. In addition to the extension, each of these models could include lateral (strike-slip) motion between the blocks of dark terrain on either side.

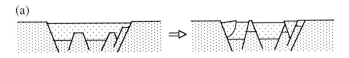

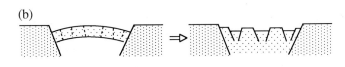

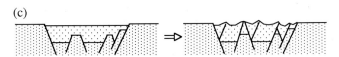

Figure 6.13. Some of the ways to account for the formation of ridges and grooves within belts of bright terrain on Ganymede, each beginning with the flooding of a fault-bounded valley. (a) Ridges and grooves formed by differential subsidence and rotation of blocks. (b) Upward bending of the surface caused by expansion as the ice freezes gives rise to extensional cracks, which, after subsidence, produce a ridge and groove topography. (c) Late-stage extrusion of viscous melt along fault-controlled fissures to construct ridges cryovolcanically.

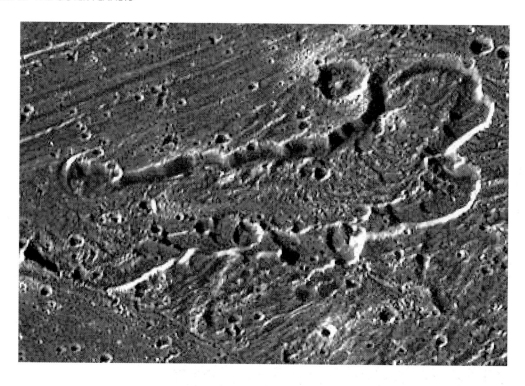

Figure 6.14. Oblique view, with south toward the top, of a 80 by 40 km region within the bright terrain of Sippar Sulcus (southeast of the area shown in fig. 6.8) seen by *Galileo*. A depression with scalloped walls opens toward the lower left and is partly filled by a flowlike deposit whose surface has ridges in a manner suggesting flow downhill from upper right to lower left.

terrain units of different ages transected by tectonic features, mainly in the form of broad fault-bounded valleys. We shall continue our survey with Dione and Tethys, two of the moons of Saturn.

6.2 DIONE

Among Saturn's moons, Dione (fig. 6.16) exhibits the most abundant signs of internal activity, with the exception of Enceladus, which is described in chapter 7. Dione is near the middle of the size range but has the highest well-determined density of Saturn's satellites apart from Titan (table 1.1). Like Rhea, it has a bright icy surface (albedo 0.5), and its leading hemisphere is distinctly brighter than its trailing hemisphere, the opposite situation to that found on Iapetus. The best images we have are at a resolution of no better than 2 km, but the global coverage is rather more complete than for Rhea.

6.2.1 Terrain Types and Tectonic Features

Three major terrain types have been identified on Dione (fig. 6.17): cratered terrain (ct) with relatively numerous craters in excess of 20 km diameter, cratered plains (cp) with an intermediate crater density, and smooth plains (sp) with a low crater density. The boundaries between these terrain units are indistinct and are not marked by any major surface structures, but they can usually be defined to within a crater width or so. The cratered terrain is significantly poorer in craters more than 20 km in diameter than even the least densely cratered regions on

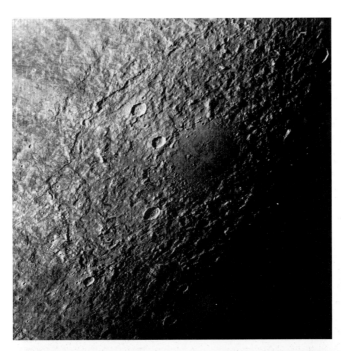

Figure 6.15. A 650 km wide *Voyager-2* image showing Gilgamesh, a young, morphologically fresh large impact structure on Ganymede. The central smooth area, toward the upper right of this view, is about 150 km in diameter. The most conspicuous concentric ring, which is 250–300 km from the center of the basin, is marked by an irregular inward-facing scarp up to 1.5 km high. Most of the craters in the lower left are secondary, having been produced by ejecta thrown out by the main impact.

Figure 6.16. *Voyager-1* mozaic covering the Saturn-facing part of the leading hemisphere of Dione. The most prominent trough in the upper left is Latium Chasma, and the largest crater in this view, Aeneas, has a diameter of about 200 km. Palatine Chasma (see fig. 6.19) is visible near the southernmost limb.

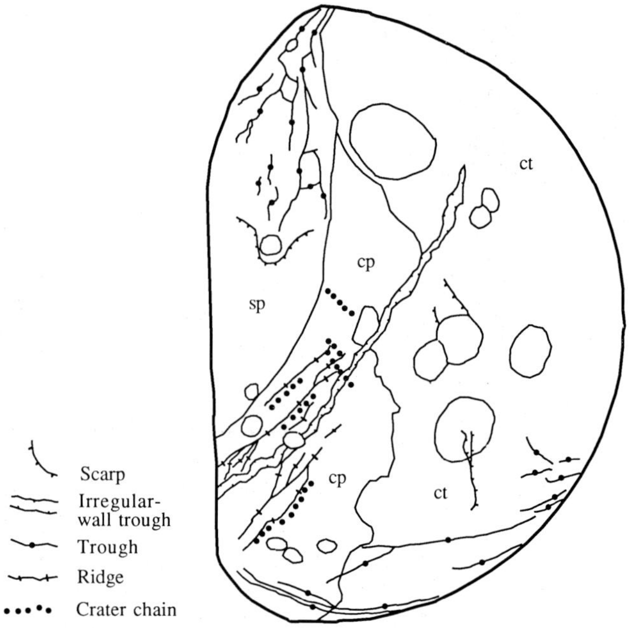

Figure 6.17. Terrain units and major landforms on Dione. The map covers the same area as the image in figure 6.16. In order of decreasing crater density, the terrain units are *ct*, cratered terrain; *cp*, cratered plains; and *sp*, smooth plains.

Scarp
Irregular-wall trough
Trough
Ridge
Crater chain

Rhea, and the variation in crater density among Dione's terrain units is much greater than the differences among regions on Rhea. It appears that the heavily cratered terrain must have been resurfaced toward the end of Population I bombardment and that the two younger terrain units represent resurfacing episodes that occurred during Population II times.

Several kinds of tectonic feature can be mapped on Dione. There are pale wispy streaks in the poorly imaged regions (fig. 6.18), some of which can be traced into areas imaged at higher resolution, where they are shown to be troughs about 8 km in width, such as Palatine Chasma (fig. 6.19a). These troughs are of a variety of ages, as can be seen by their variable overprinting relationships with craters. Some rather degraded troughs can be made out in the cratered terrain and cratered plains, and better preserved troughs are visible in parts of all three terrain units. The troughs are usually regarded as grabens, and in places where two or more troughs run closely parallel to one another there appears to be a classic horst-and-graben situation, with a linear up-faulted block between two down-faulted troughs (fig. 6.19b).

Troughs similar to Palatine Chasma are common farther north, and there is also one of a rather different form, being 20–30 km wide and having irregular, scalloped walls (fig. 6.17). Some of the more sinuous narrow troughs, such as those in the smooth plains in the north of figure 6.16, can be seen to split and terminate in pit craters. There are also several chains of coalescing pit craters, which are best seen in the plains areas. These may be collapse features, in which case they can be regarded as transitional forms of the troughs. Alternatively, it has been suggested that the troughs on Dione were the sites of cryovolcanic eruptions related to the plains-forming events; if so, the pit craters are likely to have been explosive vents.

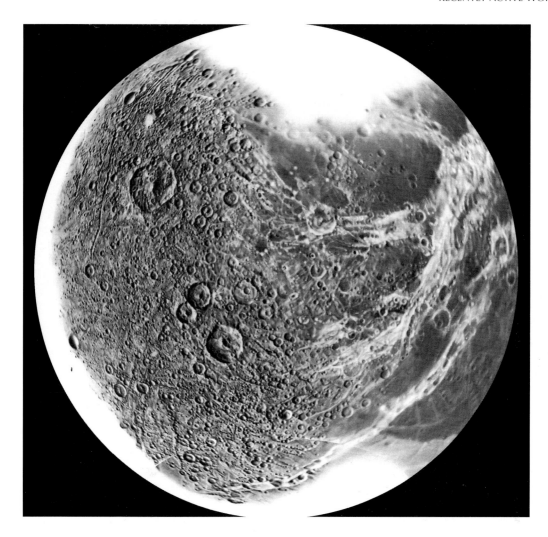

Dione also has a series of broad, convex ridges up to half a kilometer high within the cratered plains unit (fig. 6.17) that are parallel to some of the crater chains and the irregular-walled trough, suggesting a common structural control for all three types of feature. The ridge morphology could be volcanic (similar to wrinkle ridges on the lunar maria, which reflect compression caused by down-warping of the underlying lithosphere in response to the load added when a thick pile of lava has been erupted) but could alternatively indicate tectonic compression. Evidence of compression in Dione's history is also provided by a few scarps up to 100 km long, notably in the heavily cratered terrain, that are mostly linear and less than 1 km high.

Figure 6.18. Shaded relief map of Dione's Saturn-facing hemisphere. Some of the pale wispy streaks in the poorly imaged trailing hemisphere (*right*) can be traced into troughs in the leading hemisphere (*left*).

6.2.2 Geological History

It has to be admitted that the evidence we have for describing Dione's geological history is far from adequate; there are no clear signs of cryovolcanism, each of

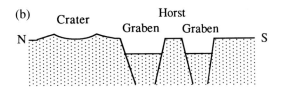

Figure 6.19. (*a*) Part of a shaded relief map of Dione's south polar region showing Palatine Chasma, a relatively fresh-looking trough that can be traced for over 600 km (see fig. 6.18). Vestiges of older troughs can be seen particularly well between 330° and 0°. (*b*) A cross section through a double portion of Palatine Chasma along the 30° meridian showing a horst-and-graben interpretation of its structure. In detail, each trough appears to be U-shaped in cross-sectional profile, probably as a result of mass-wasting of the slopes.

the tectonic features described above could be attributed to several origins, and the relative ages of different features are not well constrained. However, the images from Dione can be interpreted in a way that is broadly compatible with its supposed composition and its modeled thermal history, as outlined below.

The radiogenic heat supply within Dione would probably result in a thermal history somewhat similar to that shown for Rhea in figure 5.20, with Dione's smaller mass being partly offset by the greater ratio of rock to ice implied by its density. Equivalent models for these two satellites show that convection would be at its most intense in Dione about 180 million years after accretion and in Rhea about 140 million years after accretion, but that convection would stop sooner in Dione, at about 2.7 billion years after accretion. If resurfacing occurred during the most intensive stage of convection, the later peaking of Dione's convection may be adequate to explain why its surface mostly postdates the Population I bombardment, whereas resurfacing on Rhea went on before the Population I bombardment had declined, resulting in a more heavily cratered surface with poorly preserved signs of tectonics (at least in the well-imaged areas). The declining stage of convection would have gone on under an increasingly thick lithosphere, so the more prolonged convection on Rhea would not have been able to drive any resurfacing process.

The two plains units on Dione may represent regions of heavily cratered terrain that were flooded by eruptions of an ammonia–water cryovolcanic melt (see section 4.2.3) at different times. If so, it was presumably a low-viscosity melt because it spread over a large area (probably in many stages) and left no traces of any steep flow margins. Alternatively, the eruptions could have been explosive, in which the sudden vaporization of methane or ammonia threw out clouds of icy

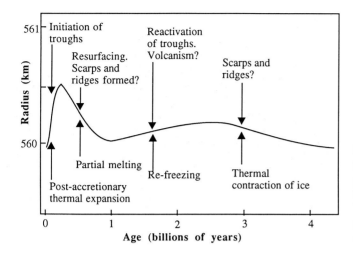

Figure 6.20. Model for changes in Dione's radius over time. The timing and amount of the initial phase of expansion and contraction are strongly dependent on the amount and relative density of the partial melt. The causes of the size changes are noted below the curve, and possible volcanic and tectonic consequences are shown above it.

particles, which would have settled to the surface as a form of volcanogenic snow. A problem with the volcanic theories is that there are no traces of flooded or buried craters showing through from the underlying surface. An alternative model is that those regions now occupied by plains units were subjected to higher heat flow, at which time all preexisting craters large enough to be seen on the images (about 2 km and above) were removed by viscous relaxation. Thus, the plains units may not represent volcanically flooded areas at all, and the various terrain ages are then due to the maintenance of anomalously warm tracts of lithosphere above regions of convective upwelling.

The timing of either the volcanic or thermal resurfacing processes would have been similar, occurring early in Dione's history when convection and partial melting were at their most vigorous and reached nearest to the surface. The extrusion of large volumes of melt would also be favored by global expansion, which would put the lithosphere under tension and thus open up pathways by which the melt could reach the surface. The results of one attempt to predict the size of Dione over time are shown in figure 6.20; unlike Rhea, Dione is too small for ice II to exist even at its center, so the only possible causes of volume change are thermal expansion or contraction, melting, freezing, and differentiation. In the model shown, the initial episode of increasing radius is due to thermal expansion of the ice as radiogenic heat built up inside. This expansion left its mark in the structure of the lithosphere, being responsible for initiating all or most of the troughs, although many of them may have been reactivated later.

An analysis of the orientation of the linear features in both the well and poorly imaged areas shows that they tend to run either to the northeast or to the northwest except in the polar regions, where they tend to run east–west. This is precisely the pattern predicted for fractures in a lithosphere where the directions of stress are controlled by tidal despinning of the satellite's rotation. It seems reasonable that Dione's linear features developed while it was being brought into captured rotation, and that the thermally predicted global expansion caused the lithosphere to fracture in directions that were dictated by the tidal despinning stress pattern.

According to this model, the result of the heating was a large amount of partial melting. As the liquid would have been denser than ice I, this led to global contraction. This could have caused some of the scarps and ridges, which seem to be compressional features. The presence of a large volume of melt would have enabled volcanic resurfacing to take place (forming the plains units) by the passage of melt up the stress-generated fractures, although the compressive nature of the stress in the lithosphere at this time, a result of the global contraction, would have hindered this. Alternatively, the strong convection facilitated by the high temperatures and the presence of the melt could have caused resurfacing by viscous relaxation. At this time, the rocky component would have had the opportunity to segregate downward to form a core perhaps 200 km in radius.

The next phase in Dione's history was when melt generation had ceased and it began to refreeze. This resulted in a prolonged episode of global expansion because of the density contrast between melt and solid, and it is possible that any volcanism that did occur went on during the early part of this phase, when melt still existed in considerable quantities and the lithosphere was once again under stress. Some troughs, especially a few that have raised rims, and chains of coalescing craters could have acted as volcanic fissures, feeding either cryovolcanic lava or explosive eruptions.

When the last of the partial melt in the asthenosphere had refrozen, the final phase of Dione's thermal history would have begun, in the form of slow global contraction caused by thermal contraction of the ice. It is unlikely that this phase has any major surface manifestations, although some of the compressional features could date from this time.

Note that the model just presented assumes that Dione as we see it has not been reaccreted after destruction by a major impact. Such an event could explain the general lack of Population I craters, but the implications of a young age are severe in greatly reducing the extent of radiogenic heating that would have been possible. That issue aside, it is almost certain that the above interpretation of Dione's history is flawed, at least in many of its details. Images with higher resolution would undoubtedly help, but we may not get significantly closer to the truth until missions have landed on Dione and have been able to examine the composition and structure of the surface material firsthand. Each of the moons of Saturn displays its own variation on the theme of thermal-related volume changes, tectonics, and resurfacing. Another permutation is seen in the next section, where Tethys, the twin of Dione, is discussed.

6.3 TETHYS

Tethys is almost identical in size to Dione, although its lower density shows that it has a smaller proportion of rocky material and so is likely to have experienced less radiogenic heating. Indeed, Tethys does exhibit fewer signs of internal activity and has elements reminiscent of Dione, Rhea, and Mimas mixed together on its surface (Figs. 6.21–6.23).

All parts of Tethys are densely cratered, much of it showing clear signs of dense Population I cratering with most of the larger craters being highly degraded by

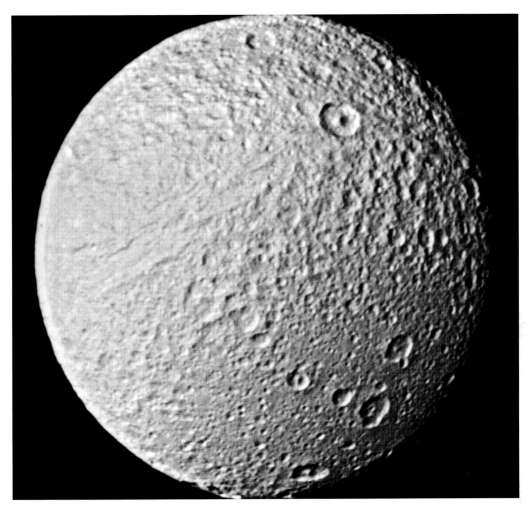

Figure 6.21. The best global view of Tethys, from *Voyager-2*, with a resolution of about 4 km. Most of the area covered is heavily cratered terrain, but the lower right of the image is occupied by a considerably less densely cratered unit (cratered plains). There are intriguing hints in the image that this terrain was emplaced by flooding. Ithaca Chasma (see fig. 6.23) runs diagonally downward to the left from the prominent crater Telemachus (*near the top right*), but is not well seen because of the high sun angle.

viscous relaxation. Part of the trailing hemisphere is covered by a lightly cratered plains unit like that on Dione, which is a reasonable sign of local resurfacing after Population I time. It seems that this resurfacing occurred by cryovolcanic flooding, because some old, larger craters have survived in this terrain but have low rims, suggesting that they were almost but not quite overwhelmed by the volcanic outpourings, whereas all the preexisting smaller craters were buried by the flood (fig. 6.21). The edge of the cratered plains unit has an intriguing appearance, but unfortunately all but one of the scheduled highest resolution (ca.

Figure 6.22. An enhanced image showing the highly relaxed crater Odysseus (*lower right*), dominating the surface of Tethys.

2 km) images of Tethys were lost (as a result of the jamming of one of the axes of *Voyager-2*'s scan platform described in chapter 3), so details such as this remain uncertain.

The surface of Tethys is marked by two giant features that make it stand out among Saturn's family of moons. The first of these is the crater Odysseus (fig. 6.22), named after the pivotal character of Homer's *Odyssey*. At a diameter of about 440 km (about 40 percent of the diameter of the satellite), this is the largest crater in the Saturnian system. Unlike the crater Herschel on Mimas, which is of only slightly smaller magnitude in relation to the world on which it occurs, the topography of Odysseus has relaxed to a large extent. Its floor has rebounded by several kilometers, and the one *Voyager* image with Odysseus on the limb shows that the floor is convex with roughly the same radius of curvature as the satellite as a whole. The crater walls are low, and the central complex is no more than a shadow of the impressive peak it must have been shortly after formation.

The second giant feature is an enormous trough, Ithaca Chasma, that extends at least three-quarters of the way round the globe (fig. 6.23). It has a width of up to 100 km and reaches about 3 km in depth. The inner walls show tantalizing evidence of terracing and other structures, but again we are thwarted by the poor resolution of the images. Poignantly, for Ithaca was Odysseus's home that he suffered such long hardship trying to return to, Ithaca Chasma falls along a great circle with the crater Odysseus lying at almost the maximum possible distance away, near one of the poles to this great circle. This has prompted the suggestion

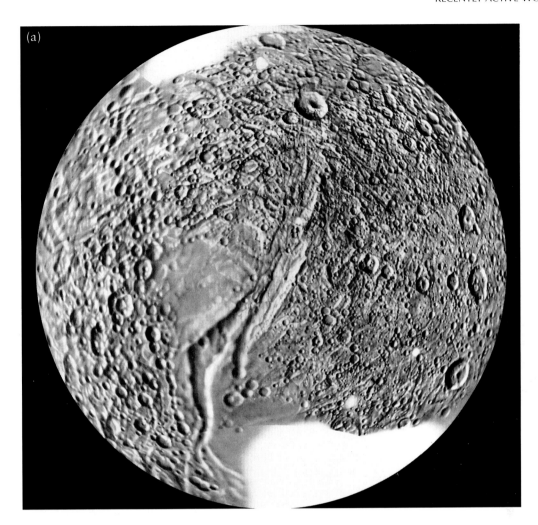

(a)

that the Odysseus impact was responsible for the formation of the chasma. The most probable mechanism is that the chasma opened up as a direct response to the viscous relaxation of the giant crater. To open the chasma in this way, the viscous flow must have been inward toward the crater *everywhere*, instead of inward in the near region but outward in regions farther away, as for major impacts whose size is less in relation to the diameter of the globe (see fig. 5.14).

In view of the size ratio between the crater Odysseus and Tethys itself, the viscous relaxation of Odysseus probes the internal structure of Tethys very deeply, and it can be shown that for the viscous flow to have been unidirectional, as suggested by the formation of the chasma, any silicate core that Tethys may possess could have occupied no more than about 20 percent of its radius at the time of relaxation. This compares with a core radius of about 47 percent if Tethys were fully differentiated. In addition, the unidirectional flow puts an upper limit on the internal thermal gradient of about 0.01 K km^{-1}. These are almost the only constraints we can put on the internal structure and differentiation of any of Sat-

Figure 6.23. Shaded relief maps of Tethys, showing the Saturn-facing hemisphere (a) and anti-Saturn hemisphere (b). Ithaca Chasma runs roughly north–south through the Saturn-facing hemisphere, and a trace of it can be seen in the northwest of the anti-Saturn hemisphere. The relief of the giant crater Odysseus is shown in an exaggerated fashion (compare with fig. 6.22).

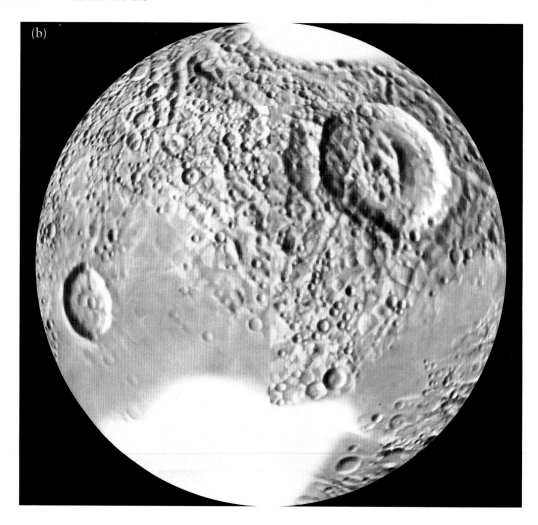

Figure 6.23. (*continued*) urn's satellites using observational evidence, and even these remain open to dispute. Even if we accept the genetic link between Ithaca Chasma and Odysseus, the lack of apparent relaxation of the chasma's topography suggests that it may have been reactivated at a later stage. Furthermore, the crater density on the floor of the chasma is slightly less than that of the cratered plains, showing that the present floor, if not the chasma as a whole, is the youngest surface yet identified on Tethys. Even so it is old, probably not significantly younger than the cratered plains units that occur on both Tethys and Dione.

We now move out to Uranus, where there are other worlds whose surfaces are marked by major troughs, with the added interest that the link between tectonism and volcanism is much clearer.

Figure 6.24. A mozaic made from the four highest resolution *Voyager-2* images of Ariel.

6.4 ARIEL

As discussed in chapter 5, Umbriel and Oberon belong in the "dead worlds" class. They are heavily cratered with few signs of other activity. Each of these moons has a "twin," in terms of size and mass, occupying a neighboring orbit of smaller radius. These are Ariel and Titania, and they bear a striking resemblance to one another rather than to their respective "twins." We begin with Ariel, which was imaged at the higher resolution, *Voyager-2*'s closest approach to Ariel being three times nearer than its closest approach to Titania.

6.4.1 Terrain Types and Tectonic Features

The best *Voyager* view of Ariel is shown in figure 6.24. This reveals a surface occupied by three distinct units, as summarized in figure 6.25: cratered terrain,

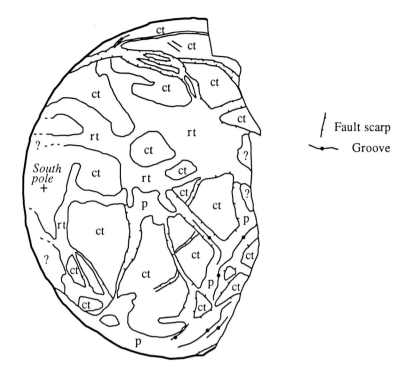

Fault scarp

Groove

Figure 6.25. Sketch map of the region of Ariel covered in figure 6.24 showing the major terrain units and tectonic features. The terrain units are *ct*, cratered terrain; *rt*, ridged terrain; and *p*, plains.

ridged terrain, and plains. The cratered terrain is evidently the oldest, though it has a lower crater density than its equivalent on any other satellite of Uranus (fig. 5.27). Some researchers attribute almost all the craters to Population II, except for a few degraded and relaxed craters in the 50–100 km diameter range that may have survived from Population I times. Others prefer to interpret Ariel's crater record as essentially Population I, with the deficit of large craters being attributed to viscous relaxation. Essentially, though, the cratered terrain appears to have been resurfaced after the peak of Population I bombardment. Within the limits of the data, there are no significant differences in age within this unit. It has completely replaced any ancient heavily cratered surface that Ariel may originally have had, such as that preserved on Umbriel and Oberon.

Areas of the cratered terrain are separated by tracts of ridged terrain, which is characterized by parallel low ridges about 10–35 km apart with shallow troughs between them. The crater density on the ridged terrain is indistinguishable from that on the cratered terrain, and so the two units evidently date from the same era. Some of the ridged terrain belts are continuous with the major tectonic features of Ariel, which are a widespread system of faults that usually bound steep-sided troughs (chasmata) from 15 to 50 km in width and 2–5 km deep. These are usually interpreted as grabens. Many of the chasmata are filled with a relatively smooth plains-forming material, which is the third of Ariel's terrain units. This plains unit provides the clearest evidence of fluid cryovolcanism encountered so far in this book, where it has spilled out beyond the end of a chasma and flooded half of a 40 km diameter crater (fig. 6.26). Where it is confined within a chasma, the plains unit commonly has a linear to sinuous medial groove in the form of a

Figure 6.26. Detail from within figure 6.24, showing an area 400 km across. The plains unit can be seen to spill out beyond the end of Sylph Chasma (which runs up the left-center from the lower edge of the image) and to flood half of an older crater, Agape, slightly above right of center. Six very subdued 10 km diameter craters can just be made out on the plains surface to the left and above Agape. Their viscously relaxed topography is best attributed to the weakness of the plains-forming material. The disruption of the right-hand wall of Sylph Chasma, 100 km above the lower edge of the image, may have been caused by a landslip.

trough 2–3 km wide. The possible significance of these troughs will be discussed shortly.

The plains unit is markedly cratered, and it evidently dates back a long time. More significantly, the crater density on it varies from place to place in a manner that suggests that the plains-forming activity continued over a substantial period. This contention is backed up by the fact that some of the fault scarps have abundant craters superposed on them, whereas others (evidently the youngest; fig. 6.27) have far fewer craters cutting them. The youngest features on Ariel appear to be several craters with bright ejecta blankets, in marked contrast to the situation on Umbriel, which, as we have seen, appears to have a uniform coat of darker material.

6.4.2 Tectonic Processes on Ariel

It has been suggested that the overall pattern of linear tectonic features on Ariel represents fractures created by tidal forces, either by tidal despinning when its rotation became synchronous, in the same way as the linear features on Dione can be accounted for, or by collapse of a tidal bulge when the orbits of Ariel and Umbriel became nonresonant (if indeed they ever were in resonance: the present ratio of their orbital periods is slightly less than 5:3). How the fracture pattern on Ariel developed its present expression, irrespective of whether or not the basic pattern reflects a tidal origin, is a matter of considerable dispute, not least because only 35 percent of the globe was imaged by *Voyager-2*. The tension indi-

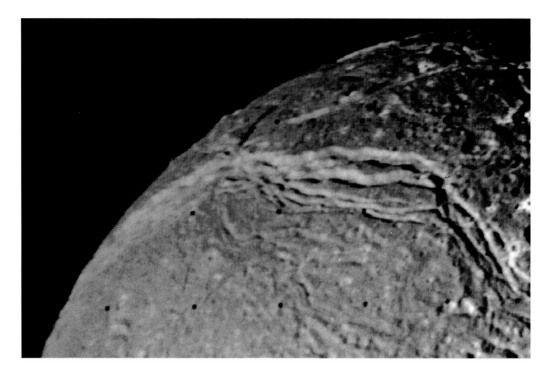

Figure 6.27. Detail from within figure 6.24 showing an oblique view of the relatively pristine horst-and-graben structure of Kachina Chasmata.

cated by the graben interpretation of the chasmata requires global expansion, unless there is a compensating amount of compression hidden in the unimaged 65 percent of the globe. Ariel is too small for internal ice phase changes, and so the most likely mechanism is the expansion that would have accompanied freezing of either water or an ammonia hydrate melt in the interior. This would have had a much greater effect on the radius of Ariel than subsequent thermal contraction of the ice.

When looked at in detail, the grabens (if such they are) on Ariel present a rather unusual aspect. The blocks of high ground separated by these grabens sometimes meet at corners, so that the direction and amount of downthrow on some of the bounding faults change instantaneously from about 2 km in one direction to about 2 km in the opposite direction. Analogs for behavior of this kind in extensional terrains on Earth and elsewhere are rare. One way to overcome this problem is to allow sideways fault movement along some of the chasmata. Figure 6.28 shows how about 50 km of lateral motion across Korrigan Chasma and Kewpie Chasma could have resulted in the corner-to-corner situation and would at the same time account for the apparent offset and deformation of Kra Chasma, which must therefore be interpreted as a preexisting structure. Alternatively, by analogy with various stepped graben systems on Earth, Kra Chasma could have formed in this stepped or "en echelon" pattern, in which case it is likely to be younger than, or roughly the same age as, Korrigan and Kewpie Chasmata. Korrigan Chasma and the offset portions of Kra Chasma are visible on the extreme right-hand edge of figure 6.26.

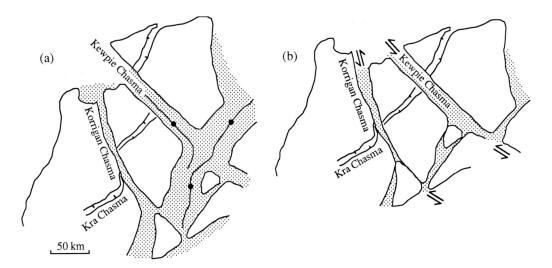

50 km

6.4.3 Volcanic Processes on Ariel

Whatever their tectonic origin and history, many of the chasmata on Ariel have clearly been flooded by some kind of cryovolcanic flow process. The area imaged in figure 6.26 provides pretty much indisputable evidence of this. However, the nature of the flow material remains debatable. A range of possible models is reviewed below.

To test their hypotheses, various researchers have tried to estimate the viscosity of the flow material on Ariel by measuring profiles across a flow. This can be done stereoscopically (using images from different viewpoints) or by photoclinometry (measuring a brightness profile and using a model of how the surface material scatters sunlight at different angles to calculate the slope). We have to assume, not always justifiably, that there are no changes in either albedo or light-scattering properties along the profile. A typical photoclinometric profile across a flow in a chasma on Ariel is shown in figure 6.29. This indicates that the flow has a strikingly convex top and that the edge of the flow is steep. The result is that the floor in the center of the chasma approaches the height of the terrain on either side and the floor is deepest at its edges, next to the walls.

One school of thought takes such profiles at face value and regards them as revealing the original morphology of the flow, in an effectively unmodified state. Working on this assumption, the steepness of the edges of the flow can be used to calculate its viscosity, which works out at about 10^{16} poise virtually irrespective of the rate at which the flow was extruded. A viscosity of this magnitude is found in water-ice at a temperature of about 240 K. This is a much lower viscosity than that of lithospheric ice in the outer solar system generally, but it nevertheless represents an extremely viscous slow-moving flow, orders of magnitude more viscous than the most viscous lavas (rhyolite) on Earth. The steep edges of the profiles would represent the fronts of the flows, because these flows would have been far too viscous to have traveled along the length of a chasma, and would have to have been erupted from fissures that ran along the chasmata floors.

Figure 6.28. Possible strike-slip faulting history of Korrigan Chasma and Kewpie Chasma. (a) Detail from figure 6.25 showing the present-day situation of blocks of cratered terrain separated by younger chasmata (stippled areas). An older chasma, Kra, appears to have been offset by sideways (strike-slip) motion across Kewpie and Korrigan Chasmata, and two blocks meet, uncharacteristically, at their corners. (b) Proposed reconstruction before the strike-slip faulting event, showing that Kra Chasma may originally have had no offsets. The double arrows show the sense of motion required to bring about the present-day situation.

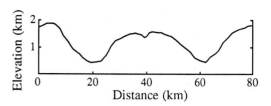

Figure 6.29. Generalized profile across a flow-filled chasma on Ariel, as derived by photoclinometry. Note the vertical exaggeration of 10:1. The example shown contains a medial groove.

There are various problems with this approach. The most fundamental from a geological point of view are how such a high temperature (170 K above the ambient surface temperature of 70 K) could be sustained near Ariel's surface, and how the heat flow from this could be localized to avoid viscous relaxation of nearby craters. For example, it has been calculated that a crater of 10 km diameter would relax completely over a timescale of 100 years if heated to around 240 K. Thus, the implied temperature makes pure water-ice an unlikely candidate for the flow material. However, it is possible to envisage other materials that would have the calculated viscosity at a lower, more feasible temperature, notably water-ice contaminated by volatiles such as ammonia, methane, nitrogen, or carbon monoxide. Traces of these in the flow material could act as lubricants for the creep of fine-grained crystalline ice at temperatures well below the 240 K required for pure ice, and this would make thermal relaxation of adjacent craters unlikely unless their ice had become similarly enriched in volatiles. In this model the subsurface ice could be mobilized by the upward migration of "warm" volatile-rich fluids, leading to extrusion of the flows through fissures parallel to (but not coincident with) the walls of the grabens. The flow-front advanced only short distances, say 10 km at most, before coming to rest. Where a medial groove occurs within a chasma, near the crest of the upwarp in the floor, it is interpreted (rather unsatisfactorily) as a junction between two flows.

An opposing school of thought notes that those viscosity models that suggest a value of 10^{16} poise are rather naive, because they ignore the fact that the likeliest cryovolcanic lavas may, like lava on Earth, have possessed a yield strength. This would mean that its viscosity would appear higher at low rates of shear strain, so it ought not to be modeled as a simple newtonian fluid. Moreover, a low-viscosity fluid can produce a flow of apparently high viscosity if it is confined by a chilled, more rigid skin or by an apron of debris shed from the advancing flow-front. When used to mimic terrestrial lava flows, comparable models commonly give viscosities that are four to six orders of magnitude too high, and so it can be reasonably claimed that Ariel's photoclinometrically derived flow profiles are compatible with flows of much lower viscosity. The most likely candidate would be an ammonia–water melt at 176 K (see chapter 4) that had largely congealed so that it consisted of more than 80 percent crystals. A crystal mush like this could have a viscosity in the range of 10^4–10^8 poise, which, allowing for Ariel's lower gravity, would behave much like a rhyolite lava flow on Earth. Similar behavior would be found if the ammonia–water mix were contaminated with a trace amount of methanol, which would lower the freezing point by about 20 K and would keep the melt viscous even in the absence of any crystals. These are attractive mechanisms in that they avoid the high heat flow necessary to mobilize pure water-ice. Medial grooves can be explained by the coalescence of two flows, as before, or as the actual fissures from which the flows were erupted.

There are also ways to account for the medial grooves as faults or fractures. The simplest tectonic explanation is that they merely demonstrate an episode of renewed faulting on the trend established in the grabens, although it is difficult to see why this should occur along new lines instead of reusing the graben-bounding faults. A refinement of this concept is to suggest that the chasma floor has been bowed upward after flow emplacement (either by viscous relaxation as in fig. 6.6 or because of intrusion beneath the floor of the chasma) and that the medial grooves are fractures caused by stretching over the center of the arch. However, if we appeal to nonvolcanic processes to explain the profiles across the grabens, the estimates of flow viscosity based on these profiles become meaningless, except as crude upper limits. We should also be wary of the profiles themselves, because we have no good grounds for assuming that the albedo and particularly the light-scattering properties of the flow material are uniform and the same as those of the older surfaces to either side.

Having accepted that the profiles across a chasma may not tell us anything about the initial steepness of the flow margins, we can consider models at the opposite extreme of viscosity from where we started. What if the flows were erupted as crystal-free ammonia–water melts? Such a fluid would have a viscosity of the order of 100 poise (about the same as honey) if crystal-free at 176 K, increasing by several orders of magnitude as crystallization progressed. The initial flow mobility could be even lower in the presence of additional volatiles. Under Ariel's low gravity and while crystal-free, an ammonia–water melt would behave in a less mobile fashion than a molten basalt on Earth or the Moon, but could be more mobile than a terrestrial flow of andesitic composition. This opens up the possibility of the flows spreading long distances from their sources, so that a flow may have traveled along the confining chasma from a point source at one end instead of oozing out sideways from a fissure running the length of the flooded region.

The images we have are not adequate to show details of any of the putative source regions on Ariel, but this mechanism is consistent with many of the observations that can be made. In the first place, it is the most satisfactory way of accounting for the way that the flows that extend beyond the ends of both Sylph and Korrigan Chasmata fan out as they escape beyond the confining walls (fig. 6.26); a chilled crust or rubble layer on the top and sides of a flow would account for the steepness of its margins, whereas the low viscosity of the interior of the flow would have enabled it to spread a long way. Second, it offers an attractive way of explaining the medial grooves as collapsed lava tubes. Tubes commonly form in basalt flows that travel many kilometers on Earth and hundreds of kilometers on the Moon. The chilled roof of the tube acts as an insulating cap that prevents significant cooling of the melt, enabling it to travel great distances before solidifying. The roof of a tube that later becomes drained of melt is liable to collapse, and this is the favored origin for most of the sinuous rilles on the Moon, some of which bear a striking resemblance to Ariel's medial grooves (fig. 6.30). There are a few examples on Ariel where a medial ridge takes the place of a medial groove. These can be interpreted as tubes that did not drain and where contraction or subsidence of the rest of the flow surface left the tube as a positive feature.

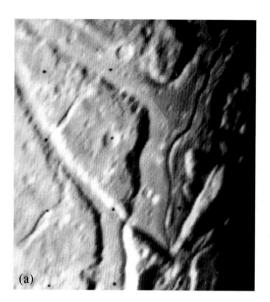

Figure 6.30. (*a*) The best *Voyager* view of Sprite Vallis, a 2–3 km wide medial groove in Brownie Chasma. This image covers much the same area as figure 6.28. (*b*) A *Lunar Orbiter* mosaic showing the sinuous rille in the lava-flooded Alpine Valley on the Moon. These two examples have comparable scales, and lend support to the concept that medial grooves on Ariel may be collapsed cryovolcanic "lava" tubes, although there are several other possible interpretations.

Perhaps it is unwise to seek a single model for the viscosity and composition of the flows on Ariel. Debates of this nature are likely to continue until we have higher resolution images covering more of Ariel's surface, although our appreciation of what might have occurred will undoubtedly improve as we learn more about the composition of the outer solar system. Furthermore, the effects of crystallinity and trace contaminants on the viscosity of low-temperature melts are only just beginning to be documented. It is probably being unreasonably optimistic to expect real flows to be pure a ammonia–water mix, and we badly need to know more about what happens when methane, methanol, and other volatiles are present in the melt. These matters are being investigated by laboratory studies of ice mixtures at low temperatures and at both low and high pressures. As this work continues, the number of possible explanations for Ariel's flows is likely to become larger rather than smaller.

6.4.4 Geological History

In view of the likely complexities of Ariel's history and its probable comparatively volatile-rich composition, convection in the solid state, if not actually involving liquids, is likely to have occurred to a sufficient extent to permit almost complete differentiation, with the possible exception of an undifferentiated crust that retains its primordial rock–ice mixture except where covered by flows. The thermal evolution of Ariel may well have followed the same lines as that of Dione, which has a similar size, density, and surface temperature. The tensional features are best explained in the same way, that is, as a result of expansion due to the onset of freezing of interior melts. We have noted that the global tectonic pattern of Ariel can be interpreted as being caused by tidal stresses. If Ariel was at times in orbital resonance with Umbriel, this would have resulted in one or more episodes of renewed interior heating, possibly involving even the melting of water, the refreezing of which would have encouraged graben formation. However, it is not certain that the relative motion across the chasmata was dominantly extensional in all instances, and some features are elegantly interpreted as being due to lateral displacements on the order of 50 km.

Tidal heating aside, Ariel's declining rate of radiogenic heat production was probably sufficient to liberate ammonia–water melts over about the first two billion years, and methane would have been mobile for even longer. The eruption of such melts would have been more likely in the earlier stages, when they could have been generated closest to the surface. Later on, the major flows are likely to have escaped to the surface along fractures that opened up under tension, hence the intimate association between the flows and the chasmata. The youngest tectonic features, Kachina Chasmata (fig. 6.27), apparently postdate the most recent episode of cryovolcanism. We shall see further evidence of eruptions and their relationship with fault systems on the remaining two major moons of Uranus, Titania and Miranda.

6.5 TITANIA

Titania is comparable in size and mass to Oberon, although in appearance it is more similar to Ariel. Unfortunately, the best *Voyager* images have a resolution of about 7 km, so we cannot compare any details. An airbrush map of Titania's southern hemisphere is shown in figure 6.31. Titania's most cratered regions have a greater crater density than the equivalent regions on Ariel (fig. 5.27), which probably means that Titania has surviving terrains that are older than anything seen on Ariel, although it is less densely cratered than Oberon and Umbriel. Titania's biggest craters appear to have a somewhat relaxed topography, notably the largest unequivocal impact structure in the Uranian system, the 275 km diameter crater Gertrude, which can be seen near the right-hand edge of figure 6.31. There are several patches with markedly lower crater densities, suggesting resurfacing probably by volcanic processes, but possibly by viscous relaxation in response to localized heating. Some areas have ridges reminiscent of those on Dione (fig. 6.17).

Titania's surface is transected by major faults forming scarps of between 2 and 5 km high and as much as 1500 km long. Opposite-facing fault scarps commonly occur in pairs or multiple sets forming 20–50 km wide grabens or horst-and-graben belts reminiscent of Kachina Chasmata on Ariel (fig. 6.27), although the temporal and spatial relationships between volcanic flooding and the grabens are less clear. These faults must be among the youngest features on Titania, because they can be seen to cut through craters and rarely have younger craters superimposed on them. Like Oberon, Titania's youngest craters are marked by bright ejecta blankets, indicating either a cleaner ice layer below the surface or a relationship between exposure age and radiation darkening. The same phenomena are indicated by the brightness of the sunward-facing walls of the grabens, which is unlikely to be due entirely to the lighting effect.

The faulting on Titania is most simply explained as a result of an episode of global expansion. A surface extension of about 1 percent seems to be indicated, which can be attributed to internal freezing in the same manner as the grabens on Ariel and Dione. If the resolution of the images were better, Titania might

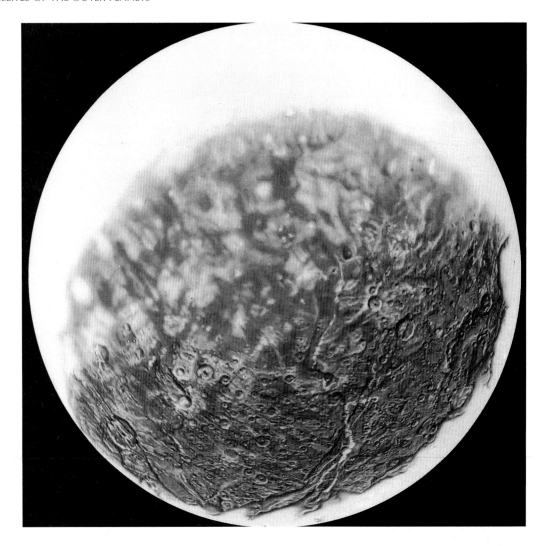

Figure 6.31. Shaded relief map of Titania's southern hemisphere. The resolution of the *Voyager-2* images was poor except in the lower (Uranus-facing) half. The faulted terrain of Messina Chasmata that extends upward from the lower edge is reminiscent of Kachina Chasmata on Ariel (fig. 6.27).

well show clearer evidence of volcanic processes. At present we cannot be sure whether or not any of the chasmata have been flooded in ways comparable to the chasmata on Ariel, though some geological maps of Titania do indicate narrow belts of smooth plains material at the foot of some of the fault scarps. However, the implications of what we can see are that Titania has been more active than Umbriel and Oberon, its neighbors on either side, and it is probably necessary to appeal to an episode of tidal heating at some time in the past to explain this activity.

6.6 MIRANDA

Miranda is the character in *The Tempest* who speaks the words "O brave new world." William Shakespeare was undoubtedly a genius in his own field, but he could scarcely have foreseen that one of the most remarkable worlds in the solar system (plate 1) would come to be named after his character.

6.6.1 Terrain Units

It was a happy chance that led to Miranda being among the best imaged bodies (in terms of resolution) in the outer solar system. *Voyager-2*'s trajectory through the Uranus system (fig. 3.7a) was chosen to take it close to Miranda in preference to any of the larger satellites simply because it was the innermost satellite known when the mission was being planned, and the spacecraft needed to pass close to Uranus to gain a gravitational assist onward toward Neptune. The mozaic of high-resolution images in figure 6.32a reveals a startlingly complex surface. About half of it has a relatively uniform albedo and a rolling topography, cut by sets of fault scarps and densely pocked by craters down to the limit of resolution. This has been interpreted to be the most densely cratered surface of all the Uranian satellites (fig. 5.27). A large proportion of these craters have a subdued appearance about their rims, quite unlike viscous relaxation and probably due to the deposition of a widespread mantling deposit of some kind. The rest of the craters are much fresher-looking (there are very few with transitional morphology) and commonly reveal a bright layer many hundreds of meters thick in their walls. The same layer can sometimes be picked out in fault scarps, and this is probably the material that forms the mantling deposit over the older craters. The other half of the surface is occupied by three regions of younger terrain of a nature not encountered elsewhere in the solar system, the interiors of which are marked by belts of bright and dark albedo material and whose margins are occupied by belts of parallel ridges. The crater density here is lower than anywhere else in the Uranian system, and the craters are all fresh in appearance. These regions have been termed *coronae* (fig. 6.32b), and each displays somewhat different characteristics.

In keeping with the Shakespearean motif, the coronae are named after places that feature in his plays. The corona at the right in figure 6.32 lies in the trailing hemisphere and is named Elsinore Corona (after the setting for *Hamlet*). This appears to be the youngest and least complex of the three coronae, although much of it is out of sight. A detail of part of it is shown in figure 6.33. The inner part, which is seen on the terminator (where the sunlit and dark hemispheres meet), is a chaotic region of presumably fault-bounded blocks, without the major albedo differences found in the other coronae (possibly an effect of the low angle of solar illumination in this region). This is surrounded by a belt of parallel ridges, each up to 1 km high and several tens of kilometers long, the outer part of which appears to be the youngest because it turns the corners at the ends of the corona (see figs. 6.33 and 6.35) and truncates the ends of ridges interior to it.

The ridges are most commonly interpreted as fracture-controlled extrusions of warm ice or a melt from among the high-viscosity options suggested for the flows on Ariel, although they might be sites of shallow intrusion, and some may reflect fault motion. The outermost ridges in places appear to be developed on top of the cratered terrain, suggesting that the ridge-forming process propagated outward from the center of the corona. The youngest feature of all in this region occurs near the top left in figure 6.33 and bears a striking similarity in morphology to a

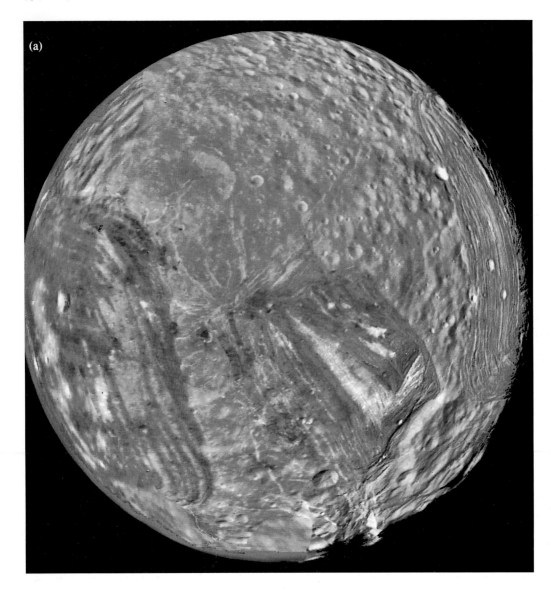

(a)

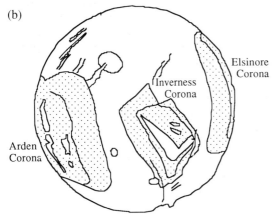

(b)

Figure 6.32. (a) Mosaic of *Voyager-2* images covering the hemisphere of Miranda that was sunlit at the time of the encounter. The Uranus-facing hemisphere is toward the bottom. (b) Sketch maps of terrains, same orientation as (a).

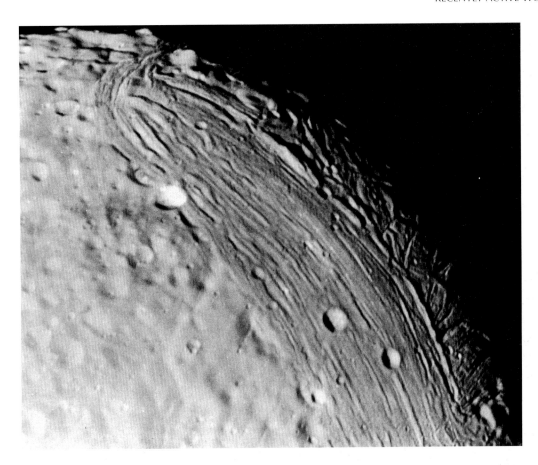

terrestrial dacite lava flow, being clearly less viscous than the material forming the ridges if they are flows as well. A sketch interpretation is shown in figure 6.34.

The corona near the center of figure 6.32 is named Inverness Corona (after the site of Macbeth's castle). It has a ridge belt around its margin comparable to that of Elsinore Corona, but its center is rather different. There is a chevron-shaped pale albedo feature, the outline of which is paralleled by a ridge and furrow pattern within it, that stops abruptly against the outer ridged belt (fig. 6.35). The whole structure is cross-cut by younger normal faults that run into the adjacent cratered terrain, one of which becomes dominant by taking over the vertical motion of several smaller faults to result in a scarp over 10 km high (fig. 6.36).

The third corona, Arden Corona (after the setting for *As You Like It*), lies on the left in figure 6.32, in Miranda's leading hemisphere. The central part of Arden Corona has bright albedo features reminiscent of Inverness Corona, although they do not form such a regular pattern, and it has a well-developed belt of parallel bands around the visible part of its periphery, giving it a superficial resemblance to Elsinore Corona. However, on closer inspection, particularly under stereoscopic scrutiny, most of Arden's bands turn out to be outward-facing fault scarps quite unlike Elsinore's smoother and more symmetric ridges. The view

Figure 6.33. Part of Elsinore Corona, with cratered terrain toward the left. Note that the craters within the corona are all sharp; there is a similar density of sharp craters in the adjacent cratered terrain, but there are also many older craters with softened edges that may have been mantled by a global event. Image about 200 km across.

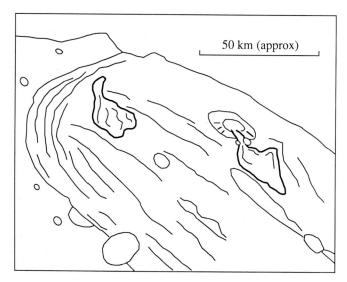

Figure 6.34. Sketch map of the upper left part of Elsinore Corona in figure 6.33 showing two probable "lava" flows (heavy outlines). The one on the left is the most obvious; the other is harder to make out, as it is partly hidden in the shadow of the ridge in the foreground.

looking across the limb of Miranda in plate 1 is of the edge of Arden Corona, near its upper right corner in figure 6.32. The dark tone of Arden's outer zone stops abruptly where it comes up against the cratered terrain, which is intensely faulted in this region. The contact with the cratered terrain makes a 90° turn here, and as it runs over the limb there is an impression of rotated fault blocks within the outer edge of the corona, as interpreted in figure 6.37.

6.6.2 Geological History

Making sense of Miranda's history is one of the more formidable problems in planetary science. An early suggestion to explain the coronae, particularly Inverness Corona with its bright chevron sandwiched between dark belts, was that Miranda as we now see it has been reaccreted following collisional breakup, and that the tonal belts are layering from the earlier body exposed end-on within large, corona-sized, reaccreted fragments. Attractive as this idea is in view of the statistical arguments that gravitational focusing of impactors would have caused Miranda to have been disrupted in this way (page 95), it does not account for the dramatic difference in crater density between the coronae and the cratered terrain. Moreover, it is difficult to imagine how any such layering could have survived with so little distortion while the reaccreted Miranda became spherical again under its own gravity. On the other hand, how could such peculiar geometries as those shown by the coronae result from essentially volcanic processes?

One crucial observation is that numerous fractures run across the face of Miranda, varying from old and subdued to younger and fresher breaks. The latter cut the coronae as well as the cratered terrain. Many of the fractures are in parallel sets with opposite-facing scarps and are most obviously interpreted as horst-and-graben structures. These represent global expansion of, at most, a few percent, which could be caused either by thermal expansion in the interior

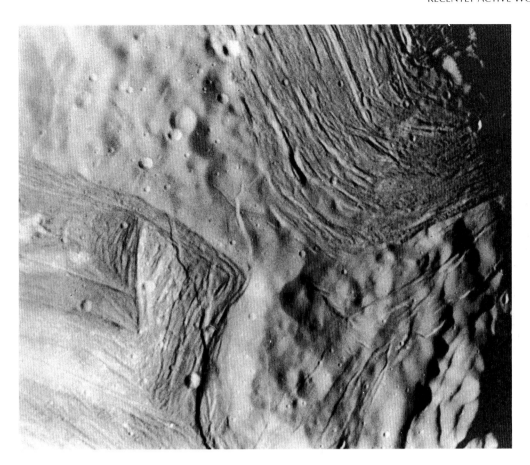

or by freezing of water. The fractures tend to run either north-northwest to east-southeast or east-northeast to west-southwest, with the same trends being reactivated several times during Miranda's history. These fracture directions are reflected in the general outlines and orientation of the ridges and tonal bands in Inverness and Arden Coronae, and to a lesser extent in Elsinore Corona, so whatever disturbances initiated the coronae, the surface manifestations were evidently controlled by the ancient fracture directions. The problem now becomes one of finding a mechanism for producing concentric structures like the coronae, taking advantage of this fracture pattern.

One interpretation is that Arden Corona occupies an impact basin, fine ejecta from which is responsible for the mantling of the topography in the ancient terrains. This basin was then filled by a series of eruptions, and maybe subsurface intrusions, of cryovolcanic melts and subsequently disturbed by reactivated tectonic fractures. Toward the end of the period of cryovolcanism in Arden, eruptions began in the region that is now Inverness Corona. Finally, after a period of quiescence, cryovolcanic melts were erupted to form Elsinore Corona, possibly reaching the surface via fractures formed antipodally during the Arden impact event.

Alternatively, the reaccretion idea can be used. Assuming that the version of Miranda that existed prior to its most recent collisional breakup was at least

Figure 6.35. Part of Inverness Corona with its pale chevron (*lower left*) and a corner of Elsinore Corona (*upper right*). Note the younger brittle fault structures especially in the lower part of the image. A bright deposit is exposed in the inner walls of the two fresh 7 km diameter pit craters between the two coronae, toward the upper left; this is probably the mantling deposit that softens the morphology of the older craters. Image 220 km across.

Figure 6.36. Detail of fault scarps near Inverness Corona, from the lower part of figure 6.32a. The major scarp on the terminator has a slope of about 60° and is over 10 km high, and there are several lower scarps parallel to it to the left, which affect the cratered terrain and the corona. The fractures immediately behind the major scarp suggest that part of it is ready to collapse in a giant landslip, and the scalloped nature of the scarp as it disappears into darkness suggests this has already happened in places. If the terrain beyond the foot of the scarp were in sunlight, deposits formed by debris avalanches might be seen.

partly differentiated, there would have been some silicate-dominated fragments among the debris so produced. On reaccretion, rocky and icy fragments would have regrouped randomly. The impact velocities would have been slow, generating too little accretional heating to allow the new Miranda to differentiate. However, if the ice was a clathrate, its conductivity could have been low enough to allow radiogenic heating to raise the interior temperature to a level sufficient to mobilize the deep interior by partial melting of ammonia–water ice. This could have been boosted by tidal heating during episodes of orbital resonance with Ariel or Umbriel, giving rise to a differentiated Miranda, with a thin icy lithosphere overlying a "warm" icy asthenosphere and a rocky core. The late accretion of a silicate-rich fragment 10–20 km in diameter now comes into play to form a corona.

This fragment could either already have been accreted in the outer part of Miranda's body, or be a piece that actually arrived during the heating event. It would arrive with low velocity (if it were derived from the breakup of the earlier Miranda), so it would not be severely disrupted upon impact. Instead, having penetrated the lithosphere, it would sink through the less dense icy asthenosphere. Figure 6.38 shows the type of flow pattern that this movement would cause. The stresses on the overlying lithosphere are compressional, and would result in chaotically oriented compressional features on the surface directly above

Figure 6.37. Sketch cross section of the edge of Arden Corona as seen in part of plate 1, interpreted as a series of rotated fault blocks. There is darker material below the original surface, which is exposed on the rotational faults.

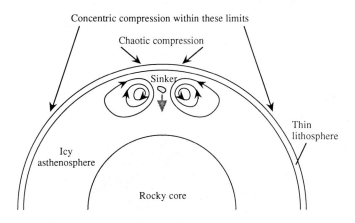

Figure 6.38. Flow in a viscous asthenosphere caused by a sinking mass, showing the resulting stresses in the overlying lithosphere. This process could initiate the formation of a corona on Miranda.

the sinking mass, changing to concentric compressional features at distances greater than 15° from the center. It has been argued that the central chaotic region of Elsinore Corona and its outer zone of concentric ridges (fig. 6.33) indicate that asthenospheric flow of this kind took place, and that the spacing of features is compatible with a lithospheric thickness in the range of 3–30 km at the time of formation. Arden Corona could have formed in the same way. However, others have used the same morphological evidence in the coronae to claim that these are sites of *uprise* of less dense material (a rock-free diapir, or pod, of icy material), which has the attraction of cryovolcanism being a natural consequence.

Whatever the origins of the coronae, their subsequent appearance has definitely been affected by later fault movements (fig. 6.37) and at least some cryovolcanism (fig. 6.34). As has already been remarked, the form of the ridges surrounding Elsinore Corona is compatible with their having been constructed by the extrusion of viscous fluids, but they could also represent linear swells above rising intrusions of less dense material. Compressional stresses do not encourage the extrusion of melts, because such stresses tend to close up any pathways to the surface. Melting caused by a corona-initiating impact is one way to produce melts in a setting where they would have had easy access to the surface.

It should now be apparent that there is no shortage of theories that seemingly could account for the coronae and various aspects of their morphology. What is still lacking is a general consensus about what really did happen. Inverness Corona is less clearly explicable on the "sinker" model than either of its companions, and it can be accounted for entirely by volcanic processes controlled by the preexisting fracture pattern, although why volcanism should be concentrated in this region remains unexplained. A geological interpretation is shown in figure 6.39. Four units are identified. On this model, unit A is stratigraphically the lowest and therefore the oldest unit in the corona. It is medium-dark and is characterized by ridges parallel to the outer edge. The edge itself is a convex lip that in

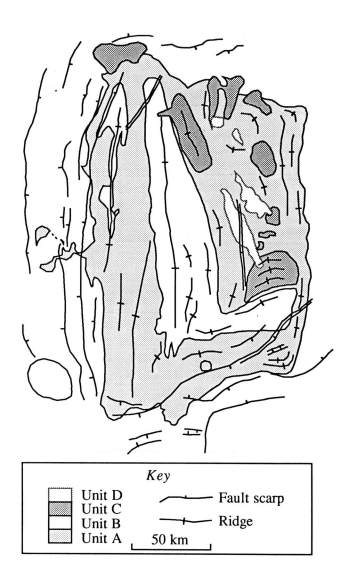

Figure 6.39. Geological interpretation of Inverness Corona, dividing it into four units, A-D. See text for explanation.

Key

▨	Unit D	⟋┤ Fault scarp
▩	Unit C	⊢┼⊣ Ridge
▤	Unit B	
▦	Unit A	├─────┤ 50 km

places has outward projections into adjacent hollows. This is consistent with an origin as a flow deposit, which is made even more clear if some dark markings beyond the western edge of the corona are interpreted as extensions of this unit spilling over the fault scarps and ponding in craters and other hollows.

Unit B forms the bright chevron in the center of Inverness Corona. Its convex lip suggests that it overlies unit A, which it resembles except for having a higher albedo. It may represent a purer ice with fewer silicate impurities, or a flow of identical composition to unit A but with a different crystal size or surface chilling history. Unit C overlies both A and B. It is slightly darker than unit A and has superimposed its own pattern of ridges over the preexisting corona topography; it appears to represent material similar to unit A but fed from localized vents rather than through fissures parallel to the overall fracture pattern. The flow units in Inverness Corona are thinner than those on Ariel (fig. 6.26), which, taking into account Miranda's lower gravity, suggests flows with a viscosity about a tenth that inferred on Ariel. This is compatible with some kind of partly crystalline ammonia–water mixture, with or without an additional volatile such as methanol. Unit D consists of bright patches with no apparent topographic effect, and is probably a dusting of pyroclastic "snow" from explosive eruptions or patches in the earlier units that became discolored as a result of some form of fumarolic activity caused by escaping gases.

Interpretations such as the above, relying on observations made at the limits of the resolution of the images, are subject to considerable dispute. However, provided they are correct in essence, if not in detail, it is apparent that volcanism on Miranda and certain other icy satellites is comparable to that encountered on Earth in terms of its complexity and variety of morphological expression.

We have now completed our survey of the recently active moons, and can turn in the next chapter to those where geological activity is continuing today. We shall begin with Io, where there is a variety of features whose volcanic origins are in no doubt whatsoever.

7 Active Worlds

O brave new world . . .
Shakespeare,
The Tempest

Four satellites are regarded as active in this book: Io and Europa (Jupiter), Enceladus (Saturn), and Triton (Neptune). The occurrence of present-day activity on Europa and Enceladus can be disputed, but there is no doubt whatsoever in the case of Io.

7.1 IO

Unlike the other satellites discussed in this book, Io does not have an icy surface (except for a sprinkling of sulfur dioxide frost). Its density and size are intermediate between those of the Moon and Mars, and in most respects it can be regarded as a terrestrial planet that happens to be in orbit around Jupiter. Jupiter's influence is manifested by the elevated temperature in the inner part of the proto-jovian nebula that seems to have prevented a substantial amount of ice from accreting onto Io. Tidal heating caused by orbital resonance with Europa keeps Io's interior hotter than it would otherwise now be in a rocky body of this size.

If interest by geologists in the moons of the outer planets can be said to have been sparked by a single event, the occasion would be the discovery of active volcanism on Io. As noted in chapter 1, the prediction that tidal heating would maintain Io's interior at a high temperature was dramatically borne out when eruption plumes were discovered on some of the *Voyager-1* images (fig. 1.7).

Clues that Io is an unusual place were already there, however. Some of the earliest Earth-based photometric observations had shown that Io is the reddest known object in the solar system, with the possible exception of Mars, and it became progressively more certain during the 1970s that Io lacks the water-ice that can be detected in the reflectance spectra of other satellites. The red color and a

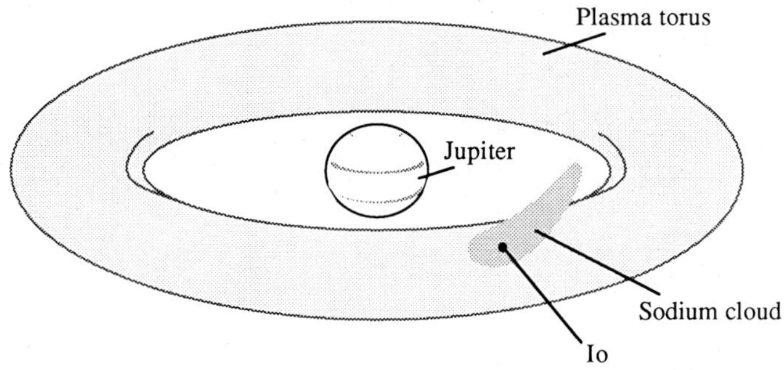

Figure 7.1. Plasma torus and neutral sodium cloud associated with Io. The cloud lies in Io's orbital plane, and the leading part of it is skewed inward toward Jupiter.

particular absorption feature in the ultraviolet led some to suggest the dominance of sulfur on Io's surface. In 1964 it was noticed that bursts of radio waves from Jupiter were correlated with Io's orbital position. Gradually, the strong interaction between Io and Jupiter's magnetic field was revealed, culminating in the 1973 flyby of *Pioneer-10* that proved that Io's orbit actually lies within the planet's magnetosphere. The discovery in the same year of an emission line in Io's spectrum caused by sodium was the first step on the road to our present knowledge that Io is surrounded by a cloud of neutral sodium, potassium, and oxygen atoms, dispersed fore and aft of Io approximately along its orbit. Shortly afterward, ionized species, notably sulfur and oxygen, were discovered in Io's vicinity, and *Voyager* data showed that these are confined by Jupiter's magnetic field into a doughnut-shaped torus that runs completely around Jupiter at the radius of Io's orbit (fig. 7.1). It is now clear that the sulfur and other atoms forming the torus and cloud have leaked away from Io's tenuous sulfur dioxide–dominated atmosphere, and that this in turn is fed by sublimation of sulfur dioxide frost from the surface and by volcanic eruption plumes.

Another clue came from telescopic thermal infrared measurements. These show that Io's surface temperature is about 130 K by day, which is a reasonable value in this part of the solar system. However, it was discovered during the 1970s that when Io passes into the shadow of Jupiter, the amount by which its temperature drops appears less when measured at shorter infrared wavelengths than at longer wavelengths. At first there was no acceptable explanation of this phenomenon, but it can now be explained by the presence of small volcanic hot spots on Io's surface occupying a continuum of high temperatures ranging down to about 200 K surrounded by larger areas at lower, background temperatures. These make only a small contribution to the measured infrared flux from Io when it is in sunlight, but become significant when the sunlight is cut off. The peak radiation from these hot spots is in the short-wavelength infrared, so when Io's temperature is calculated by measuring the infrared radiation from the whole of its disk the temperature appears higher at progressively shorter wavelengths.

7.1.1 The Surface of Io

Voyager-1 flew by Io at a distance of 20,500 km, an even closer encounter than *Voyager-2*'s flyby of Miranda. The most detailed images have a resolution of slightly less than half a kilometer. These cover only limited areas, and a smaller

proportion of the surface was imaged with a resolution of less than 2 km than for either Ganymede or Callisto (fig. 3.6). As described in chapter 3, *Galileo* was unable to obtain any super-high-resolution images of Io on its inbound close pass of Jupiter; however, at a resolution of 3–20 km it improved on the longitudinal and temporal coverage achieved by *Voyager*, and its near-infrared mapping spectrometer was able to demonstrate the ubiquity of sulfur dioxide frost virtually everywhere across the surface except at volcanic hot spots, where the temperature is too high.

Despite their limitations, the *Voyager* images were adequate to show convincingly that impact craters more than a few hundred meters across are absent on Io. For the first time in this book we have reached a world where the surface is sculpted entirely by geological forces from within (fig. 7.2 and plates 2–5). In this case the morphology and surface markings appear to be very largely the result of volcanic processes. The redness of the surface and spectroscopic data suggest that sulfur and oxides of sulfur are important at least as a coloring agent on Io in the

Figure 7.2. Shaded relief map of Io's trailing hemisphere, as it was at the time of the May 1979 *Voyager-1* encounter, showing a surface dominated by the action of volcanic processes.

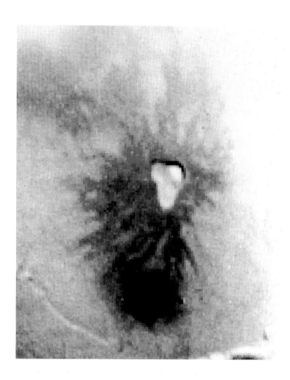

Figure 7.3. Detail of Maasaw Patera, a shield volcano on Io, from a *Voyager* image. The summit caldera is over 2 km deep, and the image is just over 100 km across.

form of either elemental sulfur or a mixture of condensed sulfur dioxide, poly-sulfur oxide, and disulfur monoxide, the latter two forming as a result of the dissociation of sulfur dioxide molecules in the high-radiation environment at Io's surface. The relative importance of sulfur volcanism versus silicate volcanism in constructing the surface remains uncertain, though undeniably at least some (and maybe most) of the volcanism involves molten rock. We shall consider some of the evidence shortly, but first we must explore the general nature of Io's surface.

The wide diversity of features on Io can be broken into three categories: vent regions, plains, and mountains. Examples of all three types are well seen in plate 2. In recognition of the discovery of active volcanism on Io, vents are named after mythological figures associated with volcanoes, fire, or thunder. Some vent regions consist of a gently sloping volcano with a steep-walled, flat-floored summit crater, or caldera, several tens of kilometers in diameter. Dark markings that look like lava flows of some kind usually radiate away from the summits. Maasaw Patera is a good example of this type of feature (fig. 7.3 and plate 2) and is reminiscent of terrestrial basaltic shield volcanoes such as in the Hawaiian and Galapagos Islands (fig. 7.4).

This sort of vent is gradational into large calderas up to about 200 km in diameter that are inset into the surface of the plains, without appearing to be at the summit of any edifice, except where lava flows can be discerned that have flowed away from the rim. The inner walls of the calderas are steep and scalloped, and the floor may occur in several levels. Such calderas (albeit Io's are on a larger scale, and generally less round) are reminiscent of terrestrial examples that are formed by collapse of the floor after a body of magma has been drained or erupted by means of lava flows or in an explosive (pyroclastic) event. The floors

Figure 7.4. A satellite image of Volcan Darwin, a basaltic shield volcano in the Galapagos Islands, which is morphologically similar to Maasaw Patera (fig. 7.3) and other shield volcanoes on Io. This image is about 35 km across. The dark areas at top right and bottom left are the sea.

of many of Io's calderas are wholly or partly covered by dark material, which is taken to be fresh lava. Examples are shown in figure 7.5.

In addition to the common types of volcanoes just described, Io has two known examples of a rather remarkable kind of volcanic construct, shown in figure 7.6. These are circular mounds about 180 km across, each with a summit crater and bounded by an apparently steep scarp running around their bases. In general form they bear at least a superficial resemblance to Olympus Mons on Mars, the largest volcano in the solar system, whose basal scarp is largely the result of landslides. No individual flows can be made out on the flanks of the examples on Io, suggesting that they were constructed by flows of a particularly fluid nature that spread symmetrically away from the vent. Alternatively, the overall shape of these volcanoes could reflect the fact that they have begun to spread under their own weight.

Each eruption plume detected on *Voyager* and *Galileo* images (of which more later) has its source either actually within a caldera or from dark linear fissures adjacent to a caldera. Infrared instruments carried by *Voyager* and *Galileo* have demonstrated that plume sources have anomalously high temperatures and revealed several other hot spots where there was apparently no plume at the time. *Voyager* images suggested that many of Io's vents were active because, apart from the plume and hot spot evidence, it is common to find pale discolored areas, possibly condensed sulfur dioxide, around either a whole caldera or just an individual flow that provides evidence of recent fumarolic activity. This was confirmed when *Galileo* began to provide images of Io seventeen years after the *Voyager* encounters, which revealed many fresh deposits, some interpreted as fallout from eruption plumes and other less extensive changes as lava flows (fig. 7.7, plate 5). There were also changes that occurred during the course of *Galileo*'s mission in the Jupiter system (fig. 7.8).

Most of Io's surface is occupied by plains with little apparent relief (fig. 7.9). The plains are often blotchy in appearance with various light and dark patches. Such color variations have been variously attributed to lava flows of different

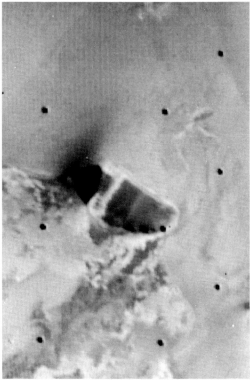

Figure 7.5. Two paterae on Io, each with dark areas on their floors that probably represent recent lava of either silicate or sulfur composition, as seen by *Voyager*. (*Left*) Creidne Patera, 160 km from north to south, flanked on either side by layered deposits associated with Euboea Montes that reach about 10 km above the plains. Creidne Patera is also visible near the center of plate 2 and was an infrared hot spot during the *Voyager-1* encounter. (*Right*) Tol-ava Patera, 80 km in length, with blotchy areas to its south suggestive of lava flows. The black dots are reseau marks.

ages or different compositions, deposits formed by condensation of gases emitted at fumaroles, and chemical alteration of the ground due to the passage of such fumarolic gases. In regions where the Sun was low enough in the sky to cast shadows, some areas can be seen to consist of superimposed layers forming tabular, smooth-topped plateaus bounded by escarpments several hundreds of meters in height. Much of the central and southern parts of the area shown in plate 2 are of this nature. On Earth, similar morphology would be explained by flat-lying layered deposits wasting away at their edges. The extent of the layers is greater than that of the lobate lava flows that can be mapped emanating from Io's vents, so if the layers are formed by flows these must be of a different kind. Fallout from eruption plumes is a possible cause of the layering; the thickness deposited at any one time ought to vary with distance from the vent, but if the force of the eruption were to change over time, the thickness of the deposit as a whole could average out to a more uniform value. An alternative explanation of the layers is that they are analogous to terrestrial ignimbrite sheets, which are formed by ground-hugging pyroclastic eruptions arising from major fissures or calderas (fig. 7.10). However, it is unclear whether on Io, in the absence of an appreciable atmosphere, the amount of gas necessary to sustain such a flow could be released by the eruption processes and remain entrained within the flow for long enough. Yet another explanation could be that the layered deposits are the result of gravity-driven debris avalanches descending from high-standing moun-

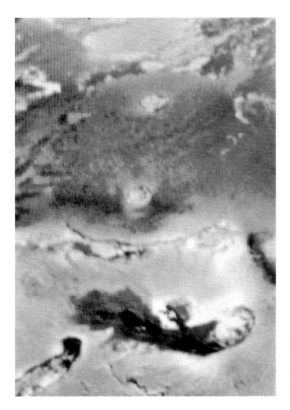

Figure 7.6. Two adjacent circular volcanic constructs on Io, Apis Tholus to the north and Inachus Tholus to the south, seen fore-shortened in this *Voyager* image. Each is about 100 km across. Their origin is enigmatic. The dark feature to their south is another example of the more common type of Io volcano, Masaya Patera.

tain blocks, which is particularly convincing around the fringes of Euboea Montes (fig. 7.5, plate 2).

None of these theories for the origin of the layering accounts for the erosion or mass-wasting that seems to have occurred since the layers were deposited. There is no wind or flowing water on Io to transport material away from the re-treating escarpments, and the lack of any obvious mechanism is one of the more unsatisfactory aspects in our understanding of Io's geology. A process that could explain the erosional modification is that the layered plateau deposits act as per-meable channels (aquifers, as we would say on Earth) for liquid sulfur dioxide, hydrogen sulfide, or some other volatile that escapes as a vapor along the escarp-ments. Bright deposits extending from the foot of some escarpments lend some credence to this view, if they are interpreted as sulfur dioxide frost, but it has been argued that such a process ought to result in the formation of explosion craters rather than the irregularly retreating continuous scarps seen on Io. An-other problem is the lack of a mechanism to transport the erosional debris away from the foot of the scarp, unless virtually all the plateau material were suffi-ciently volatile to sublime directly to vapor.

Mountains occupy less than 2 percent of Io's surface. They typically have ir-regular outlines and rugged, possibly fault- or fracture-controlled surfaces, and occur at all sizes up to 200 km in diameter and 10 km in height. The mountains are often paler and less red than the rest of Io's surface (except for the bright halo deposits) and have the appearance of protruding through the surrounding plains

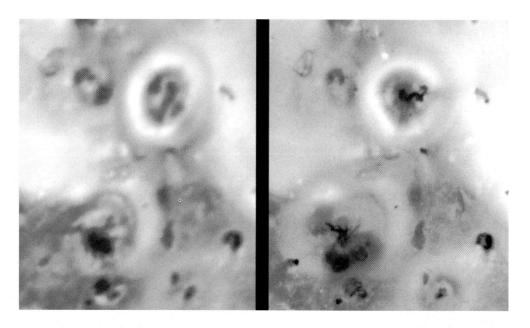

Figure 7.7. A comparison between *Voyager-1* (*left*) and *Galileo* (*right*) images showing the same 700 km wide area in 1979 and in September 1996. The bright ring toward the top of each view is the fallout from the Prometheus eruption plume. Between the dates of the two images the plume deposits have changed shape as their source migrated about 70 km west, and a presumed lava flow, 100 km in length, has flowed out toward the east. Changes are also visible in the lower left of each image around the volcano Culann Patera.

unit. Haemus Mons in plate 2 is a good example, which appears to be surrounded by a discolored zone that may result from sulfur dioxide being channeled upward along the boundary between the mountain-forming material and the more permeable plains material. The steepness and height of these massifs are both greater than could be sustained by a sulfur-dominated substrate, and it seems clear that the mountains, at least, represent exposures of Io's silicate lithosphere, whatever the nature of the plains and the associated lava flows. Possibly the mountains are uplifted because of localized compressional forces associated with the burial of successive deposits as a result of Io's continual volcanic resurfacing.

7.1.2 Eruption Plumes and Hot Spots

Before delving into the debate about the surface volcanism on Io, it is worth considering the eruption plumes. The most common type of plume persists for months to years and is usually referred to as a Prometheus-type plume, after the plume from Prometheus that was detected in every suitable *Voyager* and *Galileo* image and is the longest-lived and brightest member of its class. However, the Prometheus plume itself may be atypical because it moved by about 70 km between *Voyager* and *Galileo*, suggesting that its origin is at the end, rather than the source, of a lava flow. Generally, Prometheus-type plumes reach 50–120 km high as revealed by sunlight reflected from the tiny solid particles entrained within them, and produce pale ringlike (annular) deposits typically 200–300 km in diameter (fig. 7.7 and plate 5). Ground-based and *Galileo* infrared observations have demonstrated hot spot temperatures of at least 1000 K at their sources, although the material in the plumes themselves chills rapidly as it rises and has no hot thermal signature. Retrospective interpretation of ground-based

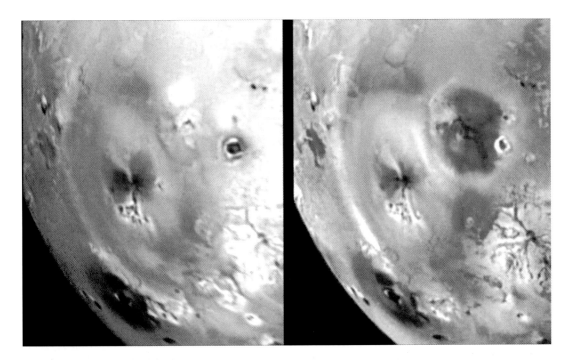

infrared observations has confirmed that Io has been volcanically active since at least the early 1970s, when the first suitable data were acquired, with a relatively steady background heat flux and temporary bursts of energy that may be related to short-lived volcanic events.

Somewhat under half of Io's plumes do not fit into the Prometheus class. A prime example is the Pele plume (plates 3 and 4, fig. 7.8), which is Io's largest. It was about 300 km high during the Voyager encounters but exceeded 400 km in 1995 and subsequent years when it was seen by the Hubble Space Telescope as well as by *Galileo*, spectroscopic measurements by the former demonstrating that the plume gas is dominantly sulfur dioxide with some sulfur. The Pele plume is fainter and presumably poorer in particles than Prometheus-type plumes, and deposits reddish materials rather than the pale materials deposited by most Prometheus-type plumes.

Prometheus-type plumes are denser than the plume observed at Pele. The plumes themselves appear dark against Io's surface (plate 5) but, as we have seen, form pale deposits (fig. 7.7). The plume shape and height, and the annular nature of the deposits, are consistent with material traveling on ballistic trajectories with exit velocities at the vent of about 0.5 km s^{-1}, perhaps from multiple vents concentrated in a source area 20 km in diameter. In contrast, the Pele plume must have an exit velocity of about 1 km s^{-1} to account for the height reached by the eruption. The Pele plume differs in form from Prometheus-type plumes by being umbrellalike in appearance, with the outer edge of the plume denser than the core (fig. 7.11).

The fissure known as Loki (fig. 7.12) has been near the site of the strongest thermal radiance from Io for most of the time since the *Voyager* encounters and

Figure 7.8. *Galileo* images of Io recorded on 4 April (*left*) and 19 September (*right*) 1997. Pele and its relatively stable plume occupy the central part of both views, but between the dates of these two images there was a high-temperature eruption at Pillan Patera (see plate 5) that distributed a 400 km wide dark deposit across the surface. Compare these with the *Voyager-1* image in plate 4 that covers roughly the same area from a different perspective.

Figure 7.9. A 1500 km wide *Galileo* image of Io's Colchis Regio, recorded in November 1996. Solar illumination is from the left. The area shown lies in the upper right of figure 7.2, the upper center of the disk in figure 7.14, and toward the top left in plate 5. There is an obvious dark lava flow near the upper left. The two dark features in the lower left and one to the right of the lava flow are known hot spots. Relief is generally subdued, but topography is evident near the terminator and below right of the lava flow where an east-facing apparent fault scarp casts a shadow.

is the source of plumes of an intermediate kind; it had a plume of Prometheus dimensions at either end during both *Voyager* encounters, but during *Voyager-2*'s approach at least one of these increased to about 330 km in height for a few days before declining to about 150 km. Usually, about 25% of Io's global heat flow emanates from Loki, making it the easiest of Io's volcanoes to observe from Earth (fig. 7.13).

Volcanic eruption plumes on Earth are driven by the explosive escape of volcanic gases (dominated by water vapor) when magma comes close to the surface. In the case of Io, the apparent ubiquity of sulfur and sulfur dioxide on its surface and surrounding it in space indicates that these, rather than water, are the most important volatile phases, and their presence has been confirmed in several cases by spectroscopic observations of the plumes themselves. Most plumes are probably driven by the escape of sulfur dioxide heated to boiling point by the intrusion of silicate magma or perhaps molten sulfur (> 393 K). Liquid sulfur dioxide, which exists between 198 and 263 K when subjected to a pressure of 1 atmosphere, has a very low viscosity (about half that of water) and so could move freely through the substrate, being drawn in from a wide area and over a long time period to replenish continuously the supply fed into a plume. It has been suggested that invisible "stealth plumes" of particle-free sulfur dioxide gas may occur from some of Io's hot spots that lack visible plumes.

Figure 7.10. Terrestrial ignimbrite flow in northern Chile, which was erupted from an explosive silicic caldera. The area shown is about 18 km across. Compare this with the tabular plateau to the northeast of Creidne Patera in figure 7.5.

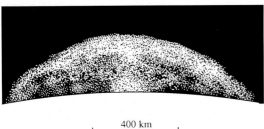

Figure 7.11. Schematic views to compare the size of Io's Pele plume (*top*) with that of a typical Prometheus-type plume (*bottom*), seen in reflected sunlight. Both are shown rising above the limb at the same scale. The Pele plume is actually considerably more diffuse and hence fainter than a Prometheus-type plume, but its appearance has been enhanced here for clarity.

7.1.3 Hot Spot Temperatures

There is clearly plenty of sulfur around on Io, and in the immediate post-*Voyager* years this prompted several people to suggest that all or most of Io's volcanism involved molten sulfur rather than molten rock. An obvious way to tell the difference is to measure the temperatures of active sites, because sulfur melts at a temperature about 1000 K lower than does rock. In fact, sulfur will begin to boil

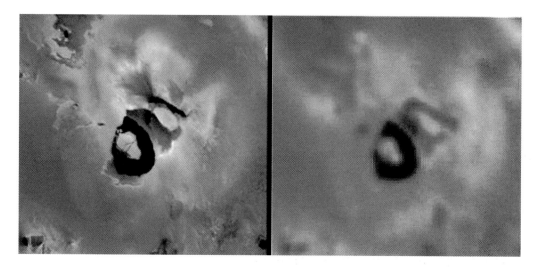

Figure 7.12. A comparison between *Voyager-1* (*left*) and *Galileo* (*right*) images showing the same 900 km wide area in 1979 and in September 1996, featuring Loki Patera. The *Voyager* image shows a dark plume emanating from the western end of the fissure on Loki's north side. There is a less obvious plume at the other end of the fissure. Loki is generally the strongest thermal feature on Io, but most of the thermal radiance appears to be from the dark "lake" to the south of the fissure. The thermal source was unusually dim in 1996, and there was no plume at the time of the *Galileo* image.

at about 700 K (or even less in the absence of any confining pressure), so temperatures in excess of this cannot indicate molten sulfur. Most early models of hot spots on Io derived temperatures in the range of 200–400 K, particularly for sites associated with Prometheus-type plumes, or 600–900 K for shorter lived events, and these data were used to suggest that Prometheus-type plumes were associated with molten sulfur volcanism. However, the derivation of temperatures of hot spots at active calderas, whether using *Voyager* or *Galileo* infrared spectroscopic data or telescopic observations from Earth, is complicated by the low resolution of the detectors (fig. 7.14). It possible to partly overcome this limitation by measuring thermal radiance simultaneously at several wavelengths, as a result of which it is now widely accepted that the true temperatures of the sources of even Prometheus-type plumes are too hot for sulfur, and exceed 1500 K in some cases.

What has allowed the data to be reinterpreted so radically is the acceptance that any volcano is certain to have surfaces at a variety of temperatures, each making a contribution to the total infrared flux. To take the simplest case of a lake of molten basalt, such as have been documented in Hawaii and elsewhere on Earth (fig. 7.15), temperatures are around 1400 K where molten lava is exposed, but most of the lake is usually covered by a chilled crust many hundreds of degrees cooler. In addition to these two volcanic components, a low-resolution sensor will also detect radiation from the area surrounding the lake that is at the normal environmental temperature, which is about 300 K for Earth and about 130 K for Io. The same arguments apply to lava flows. The only way to untangle the signals from these three sources is to measure the infrared flux at several wavelengths and fit models to the resulting curve. This is no simple matter, especially considering that in reality there are likely to be gradations between these three surface temperature components. However, if anyone were still inclined to doubt the modeled high-temperature fits to telescopic and *Galileo* near-infrared mapping spectrometer data, the matter was put beyond dispute when *Galileo*'s solid-state imaging system, operating at a wavelength of less than 1.0 µm, revealed several hot spots literally incandescent in the dark when Io was in eclipse.

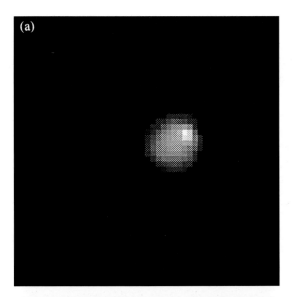

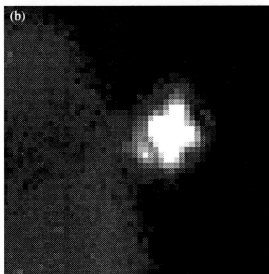

Figure 7.13. Two images of Io on 24 December 1989 from an infrared camera operating at 3.8 μm wavelength on the NASA Infrared Telescope Facility in Hawaii. (a) Io in sunlight, showing excess radiation from a source at or near Loki, appearing as a bright point toward the upper right of the disk. (b) A longer exposure image recorded after Io had passed into the shadow of Jupiter. The limb of the planet can be seen at the lower left, about to pass in front of Io. Loki is overexposed, but because the background temperature has dropped, a second, fainter, volcano can be discerned below and to the left of Loki. Its position did not match any of the hot spots observed by *Voyager*, so it was assumed to be a newly active volcano and named Kanehekili, after the Hawaiian god of thunder. *Galileo* revealed Kanehekili to be the source of a Prometheus-type plume and to consist of two hot spots about 100 km apart.

7.1.4 Silicates or Sulfur Lavas?

The evidence is now strongly in favor of the presence of molten silicates at the sources of Io's plumes and other hot spots. However, although some short-lived thermal events are best interpreted as radiation from an active lava flow, no observed hot spot can be proved to be such because the low spatial resolution of the spectrometers does not enable flows to be distinguished from the related vent. Thus, the question still remains whether Io's lava flows are Earth-like flows of molten silicate rock or, alternatively, represent cooler surface flows of liquid sulfur. The spectroscopically determined composition of the surface is no help in this respect, because it could be just a thin surface coating, unrepresentative of

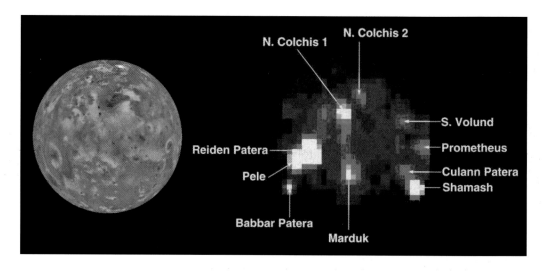

Figure 7.14. Io seen at 4.0 μm wavelength by *Galileo's* near-infrared mapping spectrometer on 7 September 1996 (right) with a *Voyager* image of the same hemisphere shown on the left. At least ten hot spots can be distinguished on this hemisphere, including that at the source of the Prometheus plume (fig. 7.7) and in the dark-floored calderas in Colchis Regio (fig. 7.9).

the underlying material. On Earth sulfurous coatings sometimes form in patches on a cooling lava flow as volcanic gases condense, and on Io this is likely to be a widespread phenomenon given the ubiquity of sulfur compounds in Io's surface environment.

The properties of liquid sulfur are rather exceptional and deserve some discussion before we can attempt to assess the evidence for molten sulfur at Io's surface. The viscosity of silicate melts is controlled largely by composition, such that silica-rich melts are typically a thousand times more viscous than silica-poor, basaltic melts. As a silicate lava flow moves, its gross composition does not change. This means that its viscosity remains fairly constant, although it will be affected slightly by decreasing temperature, the increasing proportion of crystals present, the escape of volatiles, and the confining strength of any chilled crust. In consequence, morphological differences between silicate lava flows (which may be very great) reflect their composition. Changes along the length of an individual flow (e.g., in the texture of the flow surface) are controlled by factors such as the rate of cooling and the slope of the ground. Most cryovolcanic lavas probably behave in the same way.

However, the viscosity of sulfur depends very strongly on temperature (fig. 7.16). Instead of viscosity increasing slightly as the temperature decreases (which is the case for most liquids), sulfur undergoes a drop in viscosity of a thousand-fold as the temperature falls from 440 to 430 K. The effect of this is that hot molten sulfur is rather viscous, similar to treacle or molasses, but as it cools it suddenly becomes much more fluid, comparable to hot engine oil. It has been suggested that the morphology of some of the flows on Io demonstrates just such a decrease in viscosity, notably around Ra Patera (fig. 7.17), where flows begin as very dark features around 10 km in width. These extend up to more than a hundred kilometers from the crater, becoming paler and redder, to a point where some of them suddenly fan out into broad, ill-defined orange pondlike features. The simple interpretation is that the orange ponds formed where the flow temperature had dropped below the viscosity threshold at 440–430 K and the sud-

Figure 7.15. Oblique view into the Kupianaha lava lake on Kilauea, Hawaii (March 1990). The lake was quiescent at this time, with incandescent molten material revealed only in a crescent-shaped crack at the far side of the lake. The rest of the lake's surface was covered by a chilled crust that occasionally broke apart to reveal more of the incandescent material beneath.

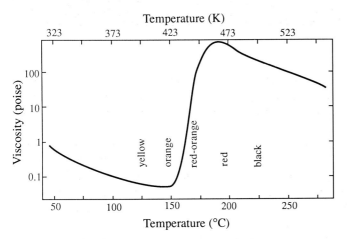

Figure 7.16. The viscosity of molten sulfur as a function of temperature, showing also the colors of liquid sulfur in each temperature range. For comparison, the viscosity of water is about 0.01 poise and the viscosity of molten basalt is about 1000 poise.

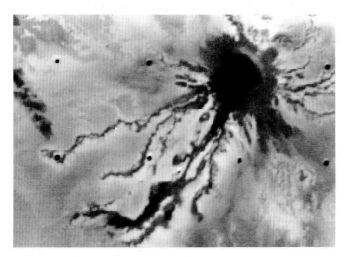

Figure 7.17. *Voyager-1* image of Ra Patera, showing flows whose morphology and color change along their length. The termination of the longest flows, which reach almost to the lower edge of this view, is marked by a broad orange pondlike feature that may represent molten sulfur becoming dramatically less viscous as its temperature fell below 440 K. The image is about 400 km across. This region looks dramatically different in lower resolution images recorded by *Galileo* in 1996 and 1997; these show that the flows heading south had disappeared, but reveal a new dark flowlike feature extending northward and several pale diffuse patches nearby.

denly decreased viscosity permitted the flow to spread out much more thinly, over a wide area. Color changes along the flows are compatible with the changes in color of molten sulfur, but even allowing for rapid quenching, there is no agreement on how these colors could survive for any length of time under Io's surface conditions, where the stable form of sulfur would be a yellow variety. The case for Ra Patera being a site of sulfur volcanism was strengthened by *Galileo*'s 1996 discovery of a diffuse 75 km high plume revealed by its ionized glow while Io was in eclipse. There was no detectable surface hot spot, but there were surface changes consistent with the emplacement of new flows and/or yellow pyroclastic material since *Voyager*'s time.

Although there is moderately good evidence in favor of the Ra Patera flows being sulfur, this is lacking in most other places. The dark material forming the floors of many of Io's calderas can be interpreted equally well as the remains of a lake of either silicate lava or sulfur. The steepness of the mountains (some of which have lava flows and small volcanic vents associated with them) and of the inner crater walls of the deeper calderas calls for a stronger material than could be formed by a succession of sulfur flows and sulfur dioxide frosts, and the high temperatures modeled in most hot-spots demonstrate that silicate volcanism is certainly widespread. Eruption of silicate lavas is thus probably more common than eruption of sulfur flows. On the other hand, the scalloped edges of the layers in some of the plains regions are hard enough to explain using a mixture of sulfur and sulfur dioxide, and seem very unlikely to occur in deposits dominated by silicates, so these parts of the surface are probably dominated by sulfur-rich material.

As for the actual composition of Io's silicate lava flows, it is likely that, as on Earth, there are many varieties. Io's flows are generally long in relation to their length, suggesting relatively low viscosities. Many flows may therefore be basalt in composition. However, modeled temperatures of 1700–2000 K in Io's hottest hot spots (e.g., at Pillan during the eruption of the new flows seen in fig. 7.8) are too high to be basalt and have led to suggestions that the molten silicates in these cases are similar to the terrestrial lavas known as komatiites, a kind of ultramafic lava poorer in silica, richer in magnesium, and lower in viscosity than basalt. Komatiites are rare today on Earth but were common before about 2.5 billion years ago when Earth's mantle was hotter. On the other hand, the repeated cycles of melting that must have been experienced by Io's crust in order to sustain anything like the present rate of volcanism for a significant fraction of Io's history make it likely that much of its crustal material is chemically strongly differentiated from its mantle, so some flows may consist of alkaline basalts richer in sodium and potassium than most basalts on Earth. There may also be flows of the rock type known as carbonatite, in which carbonates of sodium, potassium, and calcium are more abundant than silicate minerals; however, carbonatites melt at only 800 K, and as we have seen, the hot spot evidence often requires something hotter than this.

Ra Patera-type flows are exceptional, and can perhaps best be attributed to sulfur that has been melted and mobilized by the heat released by bodies of silicate melt cooling within the crust. On Earth, water vapor is the main means of transporting heat away from bodies of magma, so terrestrial analogs of Io-like sulfur flows and lakes of molten sulfur are rare. However, they do occur in ex-

ceptional circumstances (fig. 7.18). It is not yet clear whether we stand to learn more about Io by studying the dynamics of such systems, or about the terrestrial examples themselves by studying Io.

7.1.5 Internal Structure of Io

In attempting to determine Io's internal structure, a variety of factors has to be accommodated, notably measurements of Io's size, mass, gravity field and magnetic field, observations of its active volcanism, models of Io's past thermal and differentation history, and calculations of the present rate of tidal heating. Io's density (table 1.1) shows that it is essentially a rocky body. Its present waterless nature can be explained by starting with a bulk composition equivalent to carbonaceous chondrite meteorites but excluding volatiles such as methane, carbon dioxide, and water, which would be missing because of the relatively high-temperature environment close to Jupiter where Io formed, and allowing a combination of radiogenic and tidal heating to produce temperatures high enough for convection, outgassing, and ultimate loss of any volatiles that survived accretion. A silicate lithosphere would have formed, and flexing of this shell and the underlying asthenosphere would be the main means of heating by the dissipation of tidal energy. However, major gaps remain in our understanding of the depth and rate at which most of the tidal dissipation occurs.

Early models that suggested that the interior of Io is entirely molten have now

Figure 7.18. A possible analog of sulfur lakes at Io's low-temperature hot spots: ponds of molten sulfur (partly obscured by condensing water vapor) and fresh sulfur deposits around them within the crater of Poas volcano, Costa Rica (March 1989). This view shows an area about 40 m across.

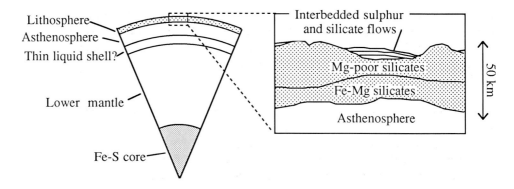

Figure 7.19. Model of the internal structure of Io. The size of the core is poorly constrained; it could extend out to half Io's radius if its iron:sulfur ratio is low and the crust is dense, or to only a quarter of Io's radius if its iron:sulfur ratio is high and the crust has low density. In the view on the left, the thicknesses of the layers above the lower mantle have been exaggerated for clarity. The detail of the lithosphere on the right is drawn approximately to scale, although its thickness is uncertain by a factor of at least two. The iron–magnesium silicate layer is analogous to the lithospheric part of Earth's mantle, and the magnesium-poor silicates and overlying flows constitute the crust. Intrusions of silicate melt into the crust probably power volcanic plumes.

been discounted, it being much more likely that heat is transported within the mantle by solid-state convection. *Galileo*'s inbound close pass by Io provided gravity and magnetic data sufficient to show that Io has a dense core (likely to be iron-rich with an indeterminate amount of sulfur), at least part of which may be molten. The core is overlain by a silicate mantle, most of which is probably solid and relatively rigid, but there may be a thin liquid shell below the lithosphere. Io's total sulfur content (about 5 percent by mass on the dehydrated chondritic model) would be sufficient to give a 50 km thick sulfur crust. Clearly, this is far more than is required to explain the nature of the surface, and some would probably remain trapped in the core. A reasonable model for the internal structure of Io, including some differentiation of the lithosphere, is illustrated in figure 7.19.

The rapid rate at which Io is being resurfaced makes it inescapable that differentiation within Io's lithosphere has resulted in formation of a distinct crust. The lack of impact craters shows that new material is added to the surface at a rate of at least 0.1 cm yr^{-1}. The heat loss from hot spots, averaged over the globe, would be equivalent to resurfacing at a rate of about 1 cm yr^{-1} if all the flows were silicate and about 10 cm yr^{-1} if all the flows were sulfur, to which plumes would add up to 1 cm yr^{-1}. At rates such as these the whole mass of Io could have been recycled many times during its history. Melting and recycling have probably been most thorough near the top of the mantle, and most models taking account of this call for the crust to be rich in the alkali metals sodium and potassium, making their abundance in Io's "sodium cloud" easier to explain. If this is the case, then the most abundant minerals in the crust would be silicates such as feldspars and related minerals such as nepheline.

Io's volcanic activity is not accompanied by clear signs of global faulting and deformation comparable with plate tectonic processes on Earth. The few faults that have been identified can be attributed to tidal flexing. If Io is in a steady state, then any fresh magma added to the crust by volcanism resulting from the rise of melts from the asthenosphere must be compensated for by material at the base of the lithosphere being removed at an equivalent rate by reincorporation into the asthenosphere. *Voyager* images suggested that volcanoes and mountains were concentrated in certain parts of the globe, but the more complete coverage achieved by *Galileo* shows that this is not the case. However, although over 500 volcanic centers have been identified on Io, during the decade ending with the

Galileo mission most of the thermal power was radiated from a small number of sites, 20–50 percent of it from Loki. The total heat flow from Io is estimated at about 1–3 W m^{-2} as a global average, compared to a global average of 0.06 W m^{-2} for Earth. This is more than can be easily accounted for by some models for Io's tidal heating, and may indicate that the process is episodic.

The widespread distribution of mountains on Io suggests that the lithosphere has no large thin patches and probably exceeds 30 km everywhere. Some of these mountains may be volcanic constructs; others have the appearance of tilted blocks that may be exposing deeper layers of the crust.

Much still remains to be understood about Io, and we are fortunate that certain things such as the cloud of neutral and ionized gases surrounding it, its spectrum and hence the composition of its surface, and the size, temperature, and location of the major hot spots can be measured and monitored from Earth. When we turn to Io's neighbor, Europa, we find a world that is equally as intriguing but that veils its mysteries in ice, giving us much less compositional information to work on.

7.2 EUROPA

With Europa, the smallest of the galilean satellites, we make a partial return to the icy worlds that predominate in the outer solar system. Europa's density is marginally less than that of our own Moon (table 1.1), but *Voyager* showed that it is not at all Moon-like to look at (fig. 7.20). It has few medium to large craters, and its appearance at low resolution is dominated by dark cracklike features. Both attributes are clear signs of young, and probably continuing, geological activity powered by tidal heating.

Europa's albedo is high and its global spectrum is dominated by fairly clean ice (fig. 2.1), though there are significant regional and local variations revealed by *Galileo* that we shall turn to later. To account for Europa's density, there must be at least 5 percent water mixed in with silicate rock. If all the water were near the surface, it would form a layer about 150 km thick, and somewhat less than this if the upper part of the silicate mantle is hydrated. Considerations of radiogenic and tidal heating, together with the paucity of impact craters and the general low relief of the surface, have been used to suggest that the lower part of the water layer is likely to be liquid. A reasonable model of the compositional layering within the outer part of Europa is shown in figure 7.21. Gravity and magnetic measurements by *Galileo* suggest that Europa has a metallic inner core occupying about 40 percent of its radius, though it is possible that Europa's magnetic signature is caused by convection within a salty ocean just below the ice rather than within the core.

7.2.1 The Voyager View of Europa

Of all the galilean satellites, Europa was the most poorly imaged by *Voyager* (fig. 3.6), but the pictures were adequate to establish the general nature of Europa's surface. We will summarize the *Voyager* view of Europa before examining the new insights provided by *Galileo*. *Voyager* images showed few craters on Eu-

Figure 7.20. *Voyager-2* view of part of Europa. Mottled terrain and plains areas are seen cut by a variety of mostly dark, narrow features. Two craters are barely visible in this image: a bright 15 km diameter crater with a dark halo (A) and, in the lower right corner, a larger multiringed crater (B). Tyre Macula, the dark spot in the upper right (C), is revealed in *Galileo* images to be a 140 km diameter crater palimpsest.

ropa, with less than a dozen between about 5 and 30 km in diameter identified with any certainty. A bright plains unit was recognized with patches of a darker (generally brownish) mottled terrain superimposed on it. Both these units were seen to be cut by narrow linear and curved bands (described as *lineae*) that are generally dark, although some, dubbed triple bands, have a bright stripe along their center; several examples of these can be seen in figure 7.20. There are also a few almost entirely bright lineae. Many lineae can be traced for several hundred kilometers, although there are also shorter, mostly younger, wedge-shaped bands that we shall turn to shortly. Superimposed on these bands and terrain types are long ridges only a few kilometers wide and about 200 m high occurring as a succession of arcs joined together (fig. 7.22) and best seen near the terminator where they are accentuated by the low sun angle. These cycloid ridges looked unlike anything else known in the solar system, but their form was taken as a strong

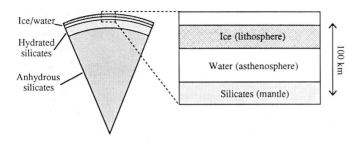

Figure 7.21. Model for the internal structure of Europa, omitting the metallic core that could occupy up to 40 percent of its radius. The extent of hydration of the silicates is uncertain, as are the relative thicknesses of the ice and liquid water layers. Other models have ice resting directly on silicates, with no intervening water layer; even so, the lower part of the ice is likely to be warm and to have a viscosity sufficiently low for it to act as an asthenosphere.

indication of compressional events, whereas the dark bands were (and remain) hard to interpret convincingly except as extensional features.

Europa is remarkable in its extraordinary low relief, and its surface is clearly one of the youngest in the solar system. With the exception of the already bright centers of the triple bands and the bright lineae, it was apparent that Europa's surface gets paler with age. This is the converse of the situation on Ganymede, but could reflect the fact that Ganymede's surface records a much longer time-scale rather than being a sign of different processes at work. The density of craters with diameters of more than 10 km on Europa is way below that seen on the lunar maria, and yet gravitational focusing by Jupiter makes it likely that Europa has experienced a greater flux of impacting bodies than the Moon over recent times. Building on this logic, the best estimate is that the average age of Europa's surface is no greater than ten to a hundred million years, but there is a wide margin for error. However, craters must be being removed by resurfacing (with help from viscous relaxation) at a much greater rate than on Ganymede, and it is probable that the processes responsible continue today.

In one region of the southern hemisphere (part of which extends into the lower left of figure 7.20), the dark bands were seen to be cut by a family of shorter, wedge-shaped dark bands that demonstrate an even younger episode of activity. They do not intersect the cycloid ridges seen by *Voyager*, and so the relative ages of these two phenomena could not be established. The major bands whose length approaches 1000 km are probably best explained by a global process (such as expansion, or tidal forces if Europa's rotation is slightly nonsynchronous), but the wedge-shaped bands are an attractive example of the scale of feature that would be likely to result from breakup of a lithosphere of the thickness shown in figure 7.21, irrespective of whether the underlying asthenosphere is warm ice or liquid water.

One of the more intriguing parts of the wedge-shaped band province as seen by *Voyager* is shown in figure 7.23. Here there are several interconnected wedge-shaped bands with convincing evidence that they cause offsets on preexisting bands. An interpretation of this region is given in figure 7.24. Here there is evidence that the icy lithosphere is broken into plates of the order of 200 km across, the relative motions between which are shown by the shape and configuration of the wedge-shaped bands separating them. Those bands that are truly wedge-shaped provide evidence of relative rotation between plates, and the amount of this rotation is in agreement with the amount of displacement experienced by the older bands that they cut. Furthermore, segments of wedge-shaped

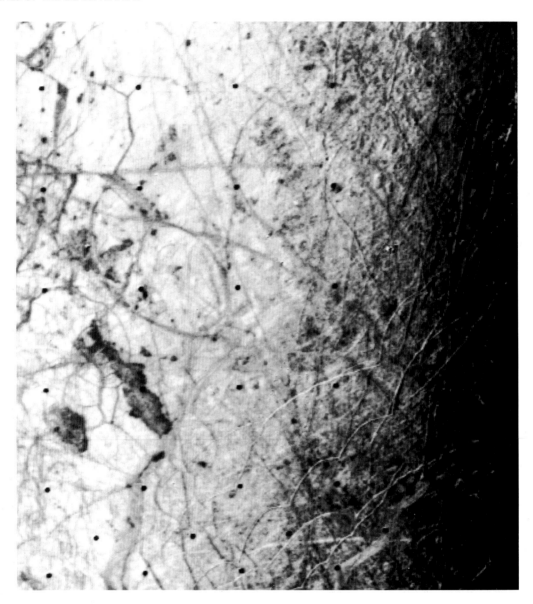

Figure 7.22. *Voyager-2* view of part of Europa's southern hemisphere, about 1800 km across and centered at about 35° south and 150° longitude. Cycloid ridges are well seen in the lower right, and a variety of irregular pits and domes can be made out in the mottled terrain toward the upper right. Dark wedge-shaped bands occur in the upper left.

band can be seen to be offset by narrow features that appear analogous to oceanic transform faults on Earth.

It should be emphasized that even if this interpretation is correct, Europa does *not* show Earth-style plate tectonics, though it appears to have the nearest thing to it that we know of in any other planetary body. The analogy between plate *B* in figure 7.24 and the Easter Island microplate on Earth (fig. 7.25) is striking, although there is an important difference: on Europa the total amount of spreading at a single spreading axis represented by a wedge-shaped band is small, whereas once spreading has started in Earth's oceanic lithosphere it is liable to continue indefinitely.

By summing the widths of Europa's wedge-shaped bands across the wedge-

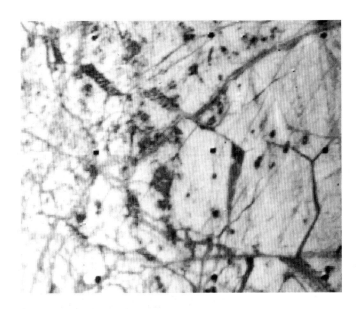

Figure 7.23. Detail of part of a *Voyager-2* image from an area overlapping the top left of figure 7.22. Wedge-shaped bands cause offsets on preexisting dark bands and a triple band. The tectonics of this region are interpreted in figure 7.24.

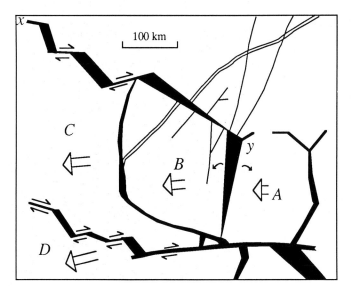

Figure 7.24. Tectonic interpretation of the area shown in figure 7.23. The sketch has been simplified by omitting some superimposed brown patches (believed, on the basis of *Galileo* data, to be salt deposits) and older fractures. Wedge-shaped bands are interpreted as marking zones of extension. The linked segments of wedge-shaped bands extending from *x* to *y* are offset by the equivalents of transform faults. Three very narrow dark bands have been displaced in a right-lateral sense across the wedge-shaped band extending northwestward from *y*, due largely to spreading by the wedge-shaped band extending south from *y*. There has been a relative rotation between plates A and B of about 5°, as implied by the southward narrowing of this feature. Note that if plate B were rotated clockwise to close up the gaps represented by the wedge-shaped bands on its northern and eastern boundaries, the displacements suffered by the narrow dark bands would be removed. In addition, plates A, B, C, and D have each migrated progressively farther to the west (relative to the northeast part of the area), as implied by the lengths of the arrows. (A triple band crossing the first wedge-shaped band shows an apparent offset, but may be a younger feature whose trend was deflected as it crossed the wedge-shaped band.)

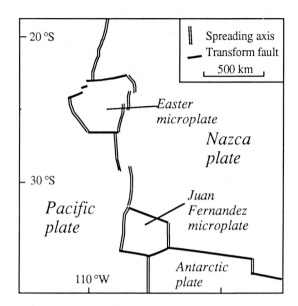

Figure 7.25. Plate tectonic map showing the spreading axes and transform faults bounding the Easter microplate in the Pacific Ocean, of which plate B in figure 7.24 is a possible europan analog.

shaped band province, the total amount of extension can be estimated at something like 100 km. Unless global expansion is allowed, there ought to be evidence of contemporaneous destruction or at least shortening of an equivalent amount of lithosphere elsewhere. Some bright bands that also appeared to be young features were suggested as possible candidates, on the grounds that, although in the *Voyager* images they did not look like terrestrial subduction zones, they could be pressure ridges of some kind. The cycloid ridges offer another possibility for compression, and parallels were drawn between them and the strings of island arcs in the western Pacific Ocean that overlie subduction zones, where the destruction of oceanic lithosphere compensates for spreading in other parts of the ocean. On Europa, however, there is no obvious geometric relationship between the directions of spreading implied by the wedge-shaped band province as a whole and the trend of the cycloid ridges.

7.2.2 The *Galileo* View of Europa

Europa was extensively imaged by *Galileo*, especially during the *Galileo Europa Mission*. There were no high-resolution images of previously known cycloid ridges, but other features and terrain types were well represented on solid-state imaging system images with resolutions as fine as 6 m per pixel (e.g., fig. 7.26), and much of the globe was mapped spectroscopically by the near-infrared mapping spectrometer at pixel sizes down to about 4 km. The spectroscopic data revealed areas where the symmetric water absorption bands diagnostic of pure ice (fig. 2.1) are distorted into asymmetric shapes characteristic of water bound in hydrated minerals. These are probably salts, and the best fits to the spectra are magnesium sulfate hexahydrite ($MgSO_4 \cdot 6H_2O$) and espomite ($MgSO_4 \cdot 7H_2O$), but carbonates of sodium such as natron ($Na_2CO_3 \cdot 10H_2O$) could be present, too. Surface salt concentration, which may reach 99% in places, is greatest along

Figure 7.26. The highest resolution *Galileo* image of Europa (13° south, 235° longitude) recorded with pixels only six meters across from a distance of only 560 km. This is an oblique view across bright plains terrain, and is only 1.8 km wide. The regular pattern of bright hills and dark valley floors in the foreground gives way to a darker region of jumbled hills beyond.

lineae such as triple bands and also in a broad region of mottled terrain shown in plate 6 and figure 7.31 that we will discuss later. The simplest explanation for these surface salt concentrations is that they are the residue from eruptions of brine onto the surface. In an explosive eruption the water could be removed from the brine by vaporization, or alternatively the salts could become concentrated on the surface during a freezing process following an effusive eruption.

Enhanced color images from the solid-state imaging system, such as plate 6, also indicate the distribution of nonice material on Europa's surface. In the wave-

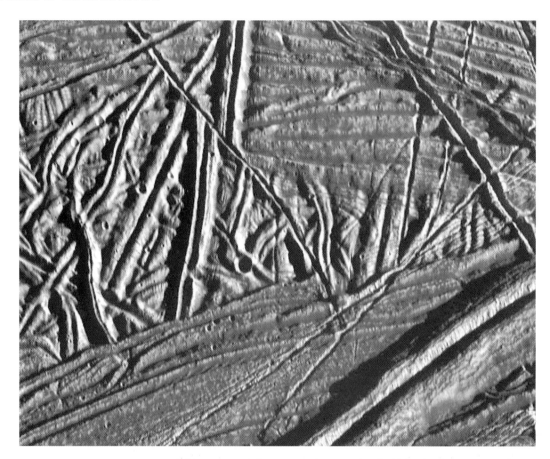

Figure 7.27. *Galileo* image, 15 km across, showing detail of a bright plains region of Europa recorded at 20 m per pixel. The double ridge cutting across the lower left is 300 m high and is one of the youngest features in the region, as can be seen from its cross-cutting relationships in plate 6, which also indicates the location of this figure. The sunlit wall of its central fissure has a fine-scale vertical fabric suggestive of upwelling and/or tearing apart, and the surface of the ridge has a texture running parallel to the central fissure indicative of incremental growth of the ridge during spreading. Note also the fracture that has caused about a kilometer of right-lateral offset to older double ridges above left of center. Impact craters from 300 m diameter down to the limit of resolution can be seen throughout.

length combination used here, clean ice is bluish gray whereas nonice, presumably salt-rich areas are progressively more red. In this region the strongest salt concentrations appear to lie along the edges of the triple bands, and their diffuse outer edges suggest explosive dispersal from eruptions along the line of the bands. Much of plate 6 covers what would have been mapped as mottled terrain at *Voyager* resolution, but the upper left belongs to the bright plains. The detail visible in plate 6 is sufficient to indicate that the plains are far from smooth, and in fact are composed of multiple ridges imparting an appearance like that of the surface of a ball of string. Seen at higher resolution (fig. 7.27), the plains are reminiscent of a fine-scale version of Ganymede's grooved terrain except that most of the more prominent ridges are effectively double, having a distinct central fissure. The morphology and texture of these double ridges strongly suggest formation by extrusion of viscous melts in an extensional setting, and it is tempting to make an analogy with midocean ridges (spreading axes) on Earth, which are elevated above the ocean floor and have central valleys. However, Europa's ridges could also be produced by explosive eruptions, and the diffuse nonice fringes to the triple bands (which are not much larger in scale than the largest double ridge in fig. 7.27) suggest that at least some material is dispersed explosively in those cases.

In a terrestrial explosive fissure eruption that begins as a curtainlike fire fountain of molten basalt along its full length, most of the fissure usually clogs up after a few hours. This confines the eruptive activity to discrete vents at sites along the fissure that develop into cinder cones. There is absolutely no sign of such localized concentration of erupted material along Europa's double ridges, which strongly suggests that they grew during episodes of extension sufficient to allow continuing eruption along the whole length of the fissure.

Galileo images revealed Europa's craters in great clarity. Tyre Macula (labeled C in figure 7.20) was revealed as a 140 km wide system of concentric circular fractures in fairly clean ice that are almost certainly the scars of an impact that fractured the lithosphere. It overprints the general texture of the surrounding plains, but it cannot be an especially young feature because it is cut by two triple bands and their associated nonice finges, which are in turn cut by several bright and apparently salt-free lineae. A much younger crater named Pwyll, whose patchy bright ejecta is superimposed on all classes of feature in plate 6, is shown in figure 7.28. Despite its evidently young age, Pwyll is viscously relaxed to the extent that its floor has risen back to the level of the surrounding plains, testifying to the thinness of the lithosphere even in comparatively recent times. Both the floor and the ejecta closest to the crater, which would have been excavated last and from deepest (fig. 5.6), are dark and probably represent a substrate from a few kilometers down that is relatively rich in nonice material.

The number of subkilometer impact craters revealed on high-resolution *Galileo* images such as figure 7.27 is suprising and at first sight runs contrary to the impression of the generally young age for Europa's surface given by the *Voyager* images. However, the flux of small impactors in the Jupiter region is particularly poorly known, and so estimates of absolute age based on the density of small craters on Europa are subject to wide margins of error. Futhermore, the density of these craters varies from place to place, suggesting that many or most could be secondary craters from a small number of relatively recent largish impacts such as Pwyll.

The region shown in figure 7.29a, which lies about a thousand kilometers northwest of the wedge-shaped bands and cycloid ridges in figure 7.22, is particularly instructive in terms of the relative ages of some of the youngest features on Europa. Both the triple bands cut, and so must be younger than, the cycloid ridge in the northeast of the image, which itself cuts a wedge-shaped band in the same corner of the image. In this region at least, the age sequence is plains formation, wedge-shaped bands, cycloid ridges, and finally triple bands. The higher resolution images in figure 7.29, b and c, showing detail from this region reveal no im-

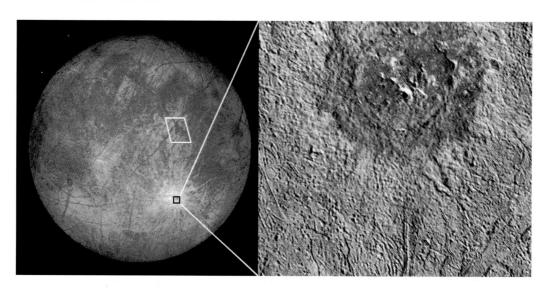

Figure 7.28. The crater Pwyll (26° south, 271° longitude) shown in a 80 km wide image recorded by *Galileo* at 250 m per pixel. Pwyll is 26 km from rim to rim. It has a dark floor and a halo of equally dark ejecta extending for about 8 km, which is presumably material excavated from several kilometers below the surface. Numerous pits at the limits of resolution are likely to be secondary craters caused by ejecta from Pwyll, and finely fragmented ejecta can be traced for more than 1000 km in the form of discontinuous bright rays, including the white patches in plate 6. The location of this image is shown on the hemispheric view of Europa on the left, which reveals the brightness of the ejecta from Pwyll. The other box on the hemispheric view indicates the location of plate 6.

pact craters on either of the triple bands, though there are some possible impact craters less than 500 m in diameter on the wedge-shaped band in (b).

Figure 7.29, a and b, reveals some significant details bearing on wedge-shaped band formation that are not apparent on *Voyager* images of similar features. As in the area shown in figure 7.23, the preexisting terrain could be reassembled by closing up the wedge-shaped bands, but in this case the resolution is adequate to reveal how the bands opened. In figure 7.29a the portion of wedge-shaped band extending northwest from the location box for (b) shows internal detail consisting of alternating lighter and darker stripes in a pattern that is symmetrical about the center of the band. This suggests that as the band opened, fresh material was added equally to the diverging plates on either side of a central fissure, the individual stripes possibly representing successive pulses of activity. The similarity with symmetric magnetic anomalies on either side of spreading axes on Earth's ocean floor is notable. However, on turning to the even higher resolution image in (b), which covers the same band just southeast of its symmetrically striped part, the wedge-shaped band is revealed to contain structure at a finer scale than previously hinted at. There is also a considerable amount of internal cross-cutting that suggests minor reorientations of short (1–10 km scale) segments of the spreading axis, which again is something familiar from Earth's ocean floor. At the scale of (b) the textural differences between the wedge-shaped band and the plains to its northeast almost disappear, and it looks as if plains (including the area shown in fig. 7.27 except for the very young double ridge in the lower right, which is both too big and too young) could be the product of multiple episodes of wedge-shaped band spreading in a variety of orientations. Some bands had reached several tens of kilometers in width (and constructed several ridges) before spreading expired, and others appeared to have died after just a single pulse of spreading had produced a solitary double ridge. If this is correct, then the wedge-shaped bands are simply the youngest, and for some reason darkest, relics of plains formation. It is well known that the smaller the particle size,

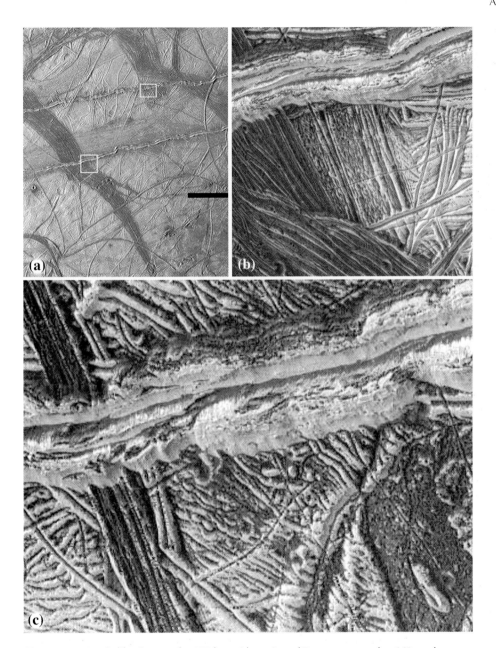

Figure 7.29. (a) *Galileo* image of a 150 km wide region of Europa centered at 16° south, 195° longitude recorded at 420 m per pixel and featuring a prominent wedge-shaped band (*lower left*) and two east–west triple bands that cut a north–south cycloid ridge in the upper right. Illumination is from the west. The black bar is missing data. Lower left and upper right boxes locate the high-resolution images (b) and (c), respectively. (b) A 20 km wide image recorded at 26 m per pixel revealing detail of the interior of the wedge-shaped band and one of the cross-cutting triple bands. Illumination is from the east. (c) A 18 km wide image recorded at 26 m per pixel, reproduced at twice the scale of (b). This shows much detail in the triple band, and a depression in the lower right that appears to have been flooded by a low-viscosity cryovolcanic fluid.

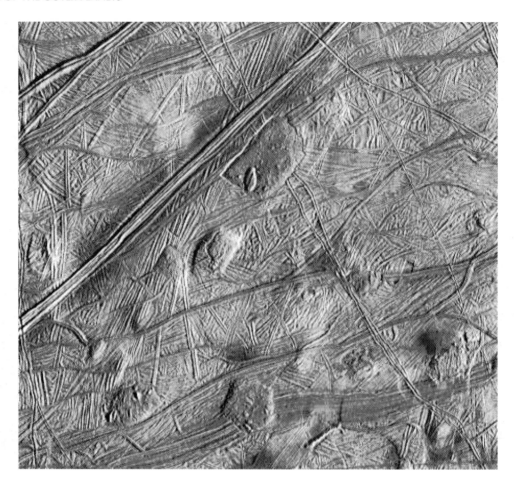

Figure 7.30. *Galileo* image of a 80 km wide region on Europa, imaged at 54 m per pixel. This is mottled terrain in which many domes have been superimposed on what was previously ordinary bright plains. Some domes are simple upwarpings of the original surface, but on others the surface has been ruptured or even completely replaced. See plate 6 for location.

the higher the albedo, so perhaps Europa's wedge-shaped band surfaces brighten with age in response to the ice crystals becoming pocked and broken by micrometeorite impacts.

Figure 7.29, b and c, reveals complexities in the morphology of triple bands that are not apparent at lower resolution. They no longer appear triple, but consist of parallel strands of discontinuous bright ridges, from which generally pale debris seems to have been shed downslope to blanket the edges of the adjacent terrain. In these monochromatic images there is no hint of the diffuse ice-poor finges extending beyond the triple bands featured in plate 6. The internal morphology of the triple bands is hard to interpret. They could be essentially compressional features, but neither these triple bands nor the cycloidlike ridges visible in the east and south of figure 7.29a show clear signs of major lithospheric shortening across them, to judge from the lack of obvious and consistent displacement of older features that they cut.

Figure 7.30 shows detail of a region of mottled terrain, which is revealed to consist of formerly normal-looking bright plains whose surface has been distorted by the development of a number of domes, ranging in size from about two to about fifteen kilometers across and up to about a hundred meters high. These

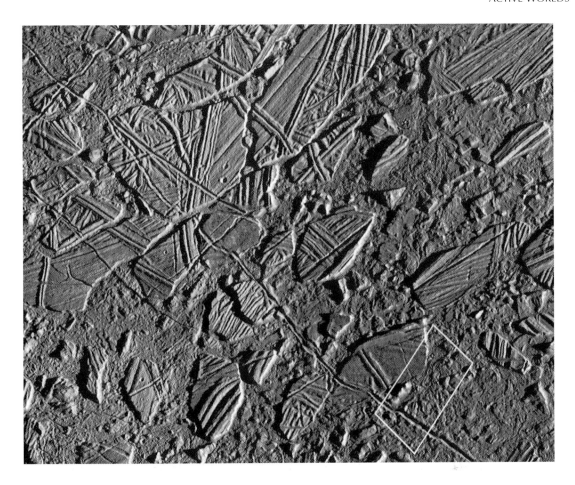

domes are obviously relatively young, and although none of them intersects the triple band running diagonally across figure 7.30, similar domes can be seen disrupting a double ridge near the bottom of plate 6, which itself cuts the triple bands near the lower right corner. It seems that these domes represent sites where pods of less dense material (described as diapirs) have risen through the crust. The material forming the diapirs is most likely to have been warm ice, made buoyant because of the thermally controlled density contrast with the surrounding colder ice, but compositional differences may also have contributed to the density contrast. The diameter of each diapir would be slightly less than that of the resulting dome. A subtle dome where the surface has retained its ridged texture would indicate a diapir having risen to about 5 km below the surface and expressed simply by upwarping of the overlying carapace, whereas a dome over which the original surface has been ruptured would indicate arrival of the diapir at a shallower depth. A few examples such as the dome in the lower center of figure 7.30 may even be viscous extrusions that have spread across the surface, either because the diapir itself reached the surface or because heat from the diapir caused local melting (or partial melting) of the overlying ice.

The dome-forming diapirs could record upward heat transport by warm ice

Figure 7.31. *Galileo* image of a 40 km wide region on Europa, showing Conamara Chaos imaged at 54 m per pixel. This is mottled terrain in which the original bright plains surface has been extensively ruptured, and is reminiscent of ice floes created by the spring breakup of the Arctic ice sheet. The outline shows the location of figure 7.32.

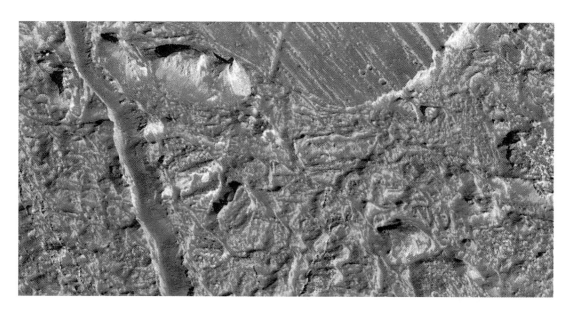

Figure 7.32. Seven kilometer wide *Galileo* image of part of Conamara Chaos recorded at 9 m per pixel. See figure 7.31 for location.

overlying a subsurface ocean. Even clearer evidence for this ocean is visible a hundred kilometers to the southwest in a part of the mottled terrain that has been named Conamara Chaos (fig. 7.31). Here the complexly ridged bright plains have become broken into a multitude of irregular fragments 1–10 km across whose edges are cliffs 100–200 m high. Between these raftlike blocks is younger material with a fine-scale hummocky texture. Plate 6 reveals that this disrupted terrain occupies a lozenge-shaped area about a hundred kilometers across. Particularly near its edges (e.g., in the northwest of fig. 7.31), the bright plains rafts have moved apart by a kilometer or less and could be fitted back together by small displacements. Toward the center of the zone there is evidence for larger movements and rotation of the rafts. Here, there must have been wholesale destruction of at least 50 percent of the old surface, and a low-viscosity fluid (now forming the hummocky interraft material) has emerged from below, either destroying the missing area by melting or causing rafts to founder. Whether this fluid was warm ice or a brine is unclear, because even had it begun as a liquid brine its surface would have frozen and become contorted into the form we now see as a consequence of continued drifting of the larger rafts. Its surface appears now to be especially salt-rich, as it lies in the heart of the region of asymmetric water absorption bands mapped by *Galileo*'s near-infrared mapping spectrometer.

Figure 7.32 shows a high-resolution view of the edge of one of the rafts and the jumbled hummocky material that surrounds it. The 250 m high hill against the edge of the raft may be the uptilted edge of a smaller raft, frozen in place in the act of foundering or tilted as the larger raft collided with it. The scale of the rafts is our most tantalizing clue as to the thickness of the solid ice shell above the supposed subsurface ocean, at least at the time when this terrain was formed. The thickness of a raft is unlikely to exceed its surface dimensions. Furthermore, assuming local isostatic equilibrium, if the intervening hummocky material on emplacement was 10 percent denser than the rafts, then the observed 100–200 m height difference between the raft surface and the surrounding area would imply a raft thickness of 1–2 km. It is difficult to imagine the density contrast being much more than 10 percent, so we seem to have a pretty reliable upper limit to the rigid ice thickness, though of course we do not know whether the immediately underlying material was a watery brine or warm, mobile ice. If the rafts were less dense than the underlying material, as implied by the fact that their surfaces are higher, then there is no reason for them to have foundered, so it is more likely that destruction of the missing percentage of old surface was by melting.

The date of the surface breakup to create Conamara Chaos can be bracketed between various events. It considerably postdates the formation of the bright plains, because small impact craters are common on these plains and their remnants as rafts, whereas there are far fewer craters on the hummocky material between the rafts. It is likely that breakup was occurring in Conamara Chaos at about the same time as diapiric action was causing the domes in most of the surrounding area. However, the new surface between the rafts must have been rigid before the renewed fracturing event responsible for the groove that runs from northwest to southeast through rafts and intervening material alike in figures 7.31 and 7.32, and as previously noted, bright ejecta from Pwyll postdates every terrain unit on plate 6.

Despite the difficulty in understanding why elements of Europa's lithosphere were so disrupted in Conamara Chaos, at least this region provides clear evidence of destruction of surface material on something like the scale required to compensate for the origin of the wedge-shaped bands (and possibly the whole of the bright plains) by incremental growth. Unfortunately, Conamara Chaos is clearly not contemporaneous with any of the wedge-shaped bands that we know of. There is another region where the bright plains surface has been rifted into Conamara-like rafts, lying at about 36° north, 86° longitude, but this appears to be similar in age to Conamara Chaos. Maybe older chaos regions are waiting to be discovered in poorly imaged regions of Europa.

7.2.3 Is Europa Still Active?

Evidently the whole of Europa's surface is younger than any large region on Ganymede, but the crater statistics do not allow us to pin down ages with any precision. An extreme interpretation calling for a rapidly declining cratering rate after 3.8 billion years ago would date diapiric intrusions and wholesale rifting events that turned Europa's bright plains into mottled terrain at about 3.0 billion years. However, more widely accepted models suggest that the likeliest age for Conamara Chaos is between a few million years and about a hundred million years. Interpretation is further complicated by the poorly known rate at which the surface ice could be eroded by bombardment by charged particles (sputtering), which might amount to 20 m per hundred million years. On top of this degree of uncertainty, we have no evidence whatsoever for the rates at which events took place. For example, the double ridge in figure 7.27 could have taken fifty thousand years to form by extension at a rate of a centimeter per year (a slow spreading rate in Earth's oceans), a decade or less if it is a viscous extrusion, or as little as a few hours if it is a spatter rampart that grew during a "fire fountaining" event at an erupting fissure. Similarly, the breakup of Conamara Chaos into rafts could have lasted millions of years or been accomplished in as little as a few weeks if it happened in the same way that Earth's Arctic pack ice breaks up every spring.

We have seen evidence that Europa's rigid ice layer can hardly have been more than a few kilometers thick under Conamara Chaos when it was breaking up, though it may have been thicker in the surrounding region that did not become disrupted. Elsewhere the sizes of the "plates" separating the wedge-shaped bands indicate lithospheric thickness of few tens of kilometers at most. It is by no means certain that the layer immediately below the rigid ice during these events was liquid; it could instead have been a soft, warm convecting ice. However, given the tidal heating expected in Europa, supplemented by radiogenic heating of the rocky interior (and even of the water/ice layer if rich in potassium salts), there was probably a liquid water (brine) layer of either local or global extent at not too great a depth below any warm ice layer. Most geologists would argue that because Europa seems to have been active within the past few hundred million years, its activity is unlikely to be finished for good. What is more in dispute is whether activity is ongoing at a uniform rate or whether it turns on and off, perhaps in response to variable tidal heating or pulses of silicate magma approaching the top of the rocky mantle.

If Europa is currently "turned off" and has been so for tens of millions of years, it is possible that any subsurface ocean is now largely frozen and the ice lithosphere has (temporarily) thickened. Be this as it may, if anything is happening on Europa today, the most convincing proof would be the discovery of an eruption in progress. Effusive eruptions are likely to be small and slow, and too cold to have a clear infrared signal. Attention has therefore focused on looking for plumes from explosive events. One *Voyager* view showing Europa as a crescent has what could be a plume rising above the limb, but the image is low quality, having been recorded at 40 km per pixel, and the suspect feature is more likely to be noise. One of the priorities for the *Galileo Europa Mission* was to hunt for plumes by imaging Europa's limb repeatedly, but nothing was found. On a geologic timescale, then, Europa is probably still active, though there have been no discernible changes over the period in which we have been observing it closely. However, if we consider what might be going on below the ice, we find that Europa could be very much alive.

7.2.4 Life on Europa?

Not too many years ago any icy satellite would have been deemed a very unlikely place to find life. However, new knowledge of the extreme environments in which microbial organisms are found on Earth shows us that bacteria and other simple organisms can survive in permanent Antarctic ice (feeding off sunlight), in 200 million year old deeply buried salt deposits (probably in a dormant state), and below a kilometer of granite (slowly metabolizing chemical energy). These environments are places where life is able to barely hang on, and suggest places to search for surviving life on planets that have become inhospitable but were formerly more Earth-like. Mars is the obvious example. However, there is one extreme environment on Earth where life is flourishing: around hydrothermal vents on the ocean floor, especially at spreading axes such as those mapped in

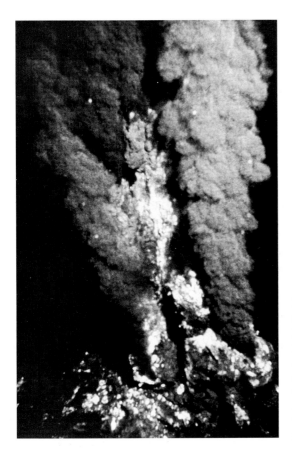

Figure 7.33. Hydrothermal vents of the type known as black smokers, seen here on a spreading axis on the floor of the eastern Pacific Ocean. Water heated by passage through underlying hot rocks, with which it has interacted chemically, escapes at about 350°C from vents about 20 cm in diameter. Upon mixing with the ambient seawater, a black precipitate of metal sulfide particles is created, giving rise to the opaque plumes. Chemosynthetic bacteria live here, deriving their energy by oxidizing the sulfides, and form the basis of an elaborate food chain that can exist independently of sunlight.

figure 7.25. The water depths here are three to four kilometers, and ambient conditions are near freezing and totally dark. Despite this, life is locally abundant because heat from igneous intrusions causes seawater to circulate convectively through fractures in the overlying crust. Cold seawater is drawn in over a diffuse outer zone, becomes heated (and therefore buoyant) as it approaches the intrusion, and is expelled upward, reaching the seabed through pipelike hydrothermal vents. During its passage through the hot rock, the circulating water picks up elements such as sulfur, iron, manganese, and barium. It thus arrives at the seafloor in a chemically enriched state, and on mixing with ambient seawater produces a plume of metal sulfide particles (fig. 7.33). Here so-called chemosynthetic bacteria make a living by oxidizing the sulfides. These bacteria in turn form the main food for a whole series of larger organisms such as clams, tubeworms, crabs, and shrimps, constituting a local ecosystem that is independent of sunlight and that would presumably continue to survive even if the sun were mysteriously removed from our solar system.

Whether there are such vigorous hydrothermal vents at the brine/rock interface on Europa depends very much on the extent of tidal heating of the rocky mantle, which is poorly known, and on whether it is still evolving by means of partial melting and silicate igneous activity (like a more subdued version of Io,

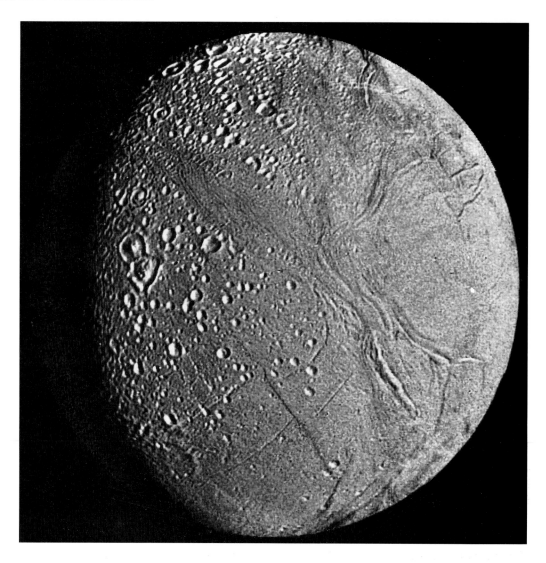

Figure 7.34. High-resolution mosaic of *Voyager-2* images of Enceladus.

buried beneath ice). However, some form of water–rock interaction seems likely. Even though such activity may be both mild and sporadic, it would almost certainly be capable of supporting the continued existence of carefully chosen microbes from Earth if we could put them there. The question then becomes: could life have already arisen on Europa? Modern thinking is that hydrothemal vents are actually the likeliest setting for life to have begun on Earth, and this is supported by genetic analysis showing that the heat-tolerant microbes found at modern hydrothermal vents are closely related to the first ancestors of all life on Earth and must be right at the base of the evolutionary tree. The indications that Europa is a likely place to harbor life are so strong that the next scheduled NASA launch to the outer solar system is a Europa orbiter (see chapter 9). If life does exist on Europa, then it probably had a chance to get started at the water/rock or ice/rock interface of several other icy satellites, too. Whether or not the Europa

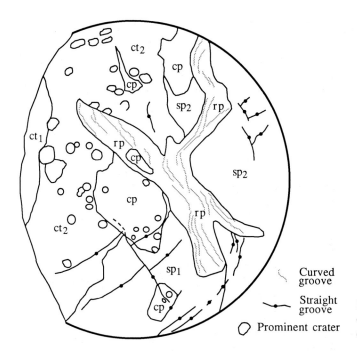

Figure 7.35. Geological sketch map of terrains on Enceladus (same area as fig. 7.34). See text for discussion.

orbiter finds direct evidence for life on Europa, the images and other data it sends back will undoubtedly add to the fascination of this remarkable world.

7.3 ENCELADUS

The closest we find to the grooved terrains of Ganymede and Europa anywhere else is on Enceladus (fig. 7.34), which has an old cratered terrain cut by and overlain by smoother units that have concentrations of grooves. This, however, is where the analogy ends, because in other ways Enceladus is a very different world from either Ganymede or Europa. For example, Enceladus is at the opposite end of the size and mass range (table 1.1) and in that respect is much closer to Mimas, but Mimas, as we saw in chapter 5, is a heavily cratered little world showing no signs of endogenic activity.

7.3.1 Terrain Units on Enceladus

Although we have useful images of less than half its surface, Enceladus displays the widest variation in crater densities and morphology of all the satellites of Saturn. Craters of comparable size may vary from fresh bowl shapes to highly degraded forms with low rims and bowed-up floors, as can be seen in figure 7.34. This is quite unlike Mimas, where there is no viscous relaxation. Various terrain unit schemes have been proposed on Enceladus, one of which is given in figure 7.35. This shows three cratered units, which, in order of decreasing age, are cratered terrain 1 (ct_1), having highly flattened craters in the 10–20 km size range; cratered terrain 2 (ct_2), having well-preserved craters in the same size range; and the cratered plains (cp), which are characterized by a lower crater density and bowl-shaped craters in the 5–10 km size range. The next youngest unit is smooth plains 1 (sp_1), which is cut by linear grooves and has a light sprinkling of craters amounting to about a thirtieth of the crater density seen in ct_1. This is succeeded by smooth plains 2 (sp_2), which is similar to sp_1 but appears to be crater-free. Around the margins of sp_2 an uncratered ridged plains unit (rp) has developed, which can be seen to have overwhelmed ct_2 along a sharp boundary where pre-existing craters are truncated.

The density of craters in ct_1, the most densely cratered unit on Enceladus, is not quite as great as that on the least-cratered areas of Saturn's other satellites. This implies an age of about 4 billion years at most, and considerably younger if the likelihood of Enceladus having been broken apart by impacts and then reac-creted is to be taken seriously (see section 5.6). Whatever the merits of that argument, the absence of craters on the sp_2 unit down to the observable limit on the best images we have (about 2 km) means that this is very unlikely to be more than a few hundred million years old.

The albedo of Enceladus is nearly uniform across all terrain units, and at virtually 1 is the highest known in the solar system. Analysis of the way the surface scatters light suggests that it is covered by fresh frost. A likely cause of this is globally dispersed spray from volcanic eruptions, perhaps associated with the generation of the sp_2 terrain. As the surface coating is uncontaminated by any detectable traces of meteoritic dust or other rocky particles, and has not become radiation darkened, this at least must be very young, and it is likely that some form of geological activity on Enceladus is either continuing today or else undergoing no more than a temporary lull.

An argument in support of this is that Saturn has a very diffuse outer ring, known as the E ring, that extends from the outer edge of the main ring system to about eight Saturn radii above the planet's surface. Photometric studies suggest that this ring consists of spherical particles about a micrometer in size, reaching a maximum concentration at an orbital radius of about 230,000 km. As table 1.1 shows, this distance coincides with the orbit of Enceladus. This suggests that Enceladus is the origin of the E ring particles, and it is tempting to attribute them to spray from explosive volcanic eruptions, just like the surface frost. Particles would not survive in such a ring for more than about ten thousand years, because they would be swept up by collision with Enceladus and other inner satellites and be destroyed by grain–grain collisions within the ring itself. This makes an apparently compelling case for continuing activity of some kind on Enceladus. However, it has been argued that, once formed, the E ring could be self-sustaining if collisions by ring particles with Enceladus threw out sufficient ejecta to replenish the ring, in which case recent activity within Enceladus would not be required.

7.3.2 Volcanic and Tectonic Processes on Enceladus

Whatever the merits of arguments for present-day activity on Enceladus, it has clearly had a prolonged active geological history, with several episodes of resurfacing. The variable degree of degradation of crater morphology within terrain units suggests that as well as changing over time, the heat flow may have varied from place to place. Perhaps the local areas of extreme crater degradation once lay above sites of convective upwelling in the asthenosphere.

Two sorts of linear or curvilinear tectonic features are distinguished on figure 7.35. The first type consists of single straight grooves that lie within both the cratered terrain and plains units. These are up to about 100 km long, and typically 2–4 km across and a few hundred meters deep. The other type of groove tends to occur in curvilinear subparallel swarms concentrated in the ridged plains unit, which lies along the edge of the sp_2 unit and in a wedge penetrating the cratered units. Curvilinear groove sets can be traced for up to 200 km.

The straight grooves are generally interpreted as tensional features. Their narrowness and the low gravity of Enceladus make it likely that they are simple extensional fractures, rather than grabens consisting of down-dropped blocks between two closely spaced faults. Enceladus is too small for ice II to exist in its interior, and so the likeliest cause of surface extension is global expansion due to the freezing of water in the interior, to form ice I, which is several percent less dense than water. There are, however, other plausible explanations for some of these grooves. One of the most striking features on Enceladus is the 20 km offset of the straight groove known as Daryabar Fossa where it crosses the straight groove Isbanir Fossa in the sp_1 terrain in the lower part of the disk in figures 7.34 and 7.35. Photoclinometric examination suggests that Isbanir Fossa is actually a 300 m high, southwest-facing scarp, whereas Daryabar Fossa has been measured as 4 km wide and more than 300 m deep where it is seen in profile crossing the limb on a *Voyager* image. It is tempting to make an analogy with an oceanic spreading axis on Earth (essentially an extensional fracture) offset by a transform fault, as shown schematically in figure 7.36a. In this model, Daryabar Fossa has acted as a spreading axis and generated the smooth plains unit within which it lies, whereas Isbanir Fossa would be a transform fault and fracture zone, formed by sideways slip of lithospheric plates generated at the spreading axis. Alternatively, Isbanir Fossa might be a conventional transcurrent fault with 20 km of right-lateral motion that caused the offset on Daryabar Fossa, which would then have to predate Isbanir Fossa (fig. 7.36b,c). This explanation is favored by the offset in the same sense and direction that Isbanir Fossa appears to cause on a second linear groove (500 m deep) within the cratered terrain 50 km to the north, but is weakened by the fact that Isbanir Fossa seems to terminate abruptly in old terrain at either end, whereas a transcurrent fault should be traceable until some related structure is reached that could accommodate the displacement.

Another explanation that has been put forward for the straight grooves is that they are volcanic fissures. This idea is favored by the existence of at least one crater chain, parallel to Isbanir Fossa, which appears to have been formed by explosive eruptions concentrated at discrete locations along a linear feature, presumably a fissure that penetrated down to a molten or partially molten zone in the interior. It is not surprising to find signs of volcanism along fractures such as this: any line of weakness in the lithosphere would be a pathway up which melts could travel. As outlined in section 4.2.3, the presence of ammonia or other volatiles in either the melt or the ice of the fissure walls could stimulate an explosive, pyroclastic eruption in several ways. Although the presence of significant amounts of volatiles in Enceladus's ice has not been independently demonstrated,

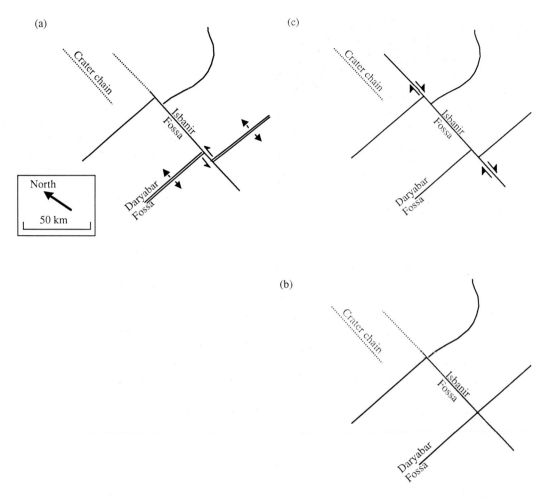

Figure 7.36. (a) Simplified tetonic map of the area including Daryabar Fossa and Isbanir Fossa on Enceladus, indicating these two fossae, an unnamed linear groove to their north, and a crater chain. The arrows show the direction of relative movement if Daryabar Fossa is interpreted as a spreading axis and Isbanir Fossa as a transform fault. (b) and (c) show two stages in an alternative model in which Isbanir Fossa is a transcurrent fault that forms after Daryabar Fossa and the unnamed groove, and causes the same offset on both. The relative age of the crater chain is not constrained by either of these models.

such eruptions are an attractive proposition because, in view of the low surface gravity of Enceladus, they could be a source for global frost deposits and even the icy droplets in Saturn's E ring. Quieter, more effusive eruptions could have led to the formation of the plains units.

The belts of curvilinear grooves, forming the young (uncratered) ridged plains unit, are another story. Each groove could represent a fissure, a narrow graben, or some other volcano–tectonic feature formed by one of the processes suggested for the grooved terrain on Ganymede, the belts of ridges in the coronae of Miranda, or even Europa's "smooth" plains. The wedge of ridged plains unit that intervenes between areas of cratered terrain toward the upper left in figures 7.34

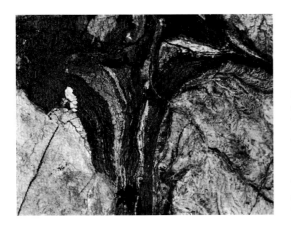

Figure 7.37. A tempting analog for the belts of ridged plains wrapping around the edges of smooth plains on Enceladus (fig. 7.34). This is a satellite image covering part of Western Australia, about 30 km across. The bright areas at the lower left and lower right are undeformed granitic cratons, and the dark terrain separating them is greenstone belts. This configuration is inherited from 2–3 billion years ago, when Earth's lithosphere was considerably thinner than today. The greenstone belts are compressed volcanic terrains that may have originated as small ocean basins.

and 7.35 consists mainly of narrow cracks and certainly has the appearance of a tensional zone. However, farther south the cracks give way to parallel 500–2000 m high rounded ridges standing above the terrain to either side that are perhaps the outer solar system's likeliest candidate for a compressionally folded terrain. This belt of ridged plains (which has been named Samarkand Sulci) wraps around the edge of the major area of smooth plains (Sarandib Planitia) visible in the right of figures 7.34 and 7.35, and continues to do so in the area out of sight in this image, beyond the upper right limb. It is tempting to see Samarkand Sulci as an example of compressional terrain deformed between more stable blocks (Sarandib Planitia on one side and the cratered plains on the other). A possible analogy is with a greenstone belt in a terrestrial granite-greenstone terrain (e.g., figure 7.37).

7.3.3 The Heat Source for Enceladus

The very young and probably continuing activity on Enceladus is in marked contrast to the situation on Dione and Tethys, which, as we saw in the previous chapter, show clear evidence of past (but not recent or continuing) geological activity. As neither of these larger and equally dense worlds shows the effects of radiogenic heating, it would be unreasonable to appeal to this as the cause of the activity on Enceladus. Models for the thermal evolution of Enceladus based both on radiogenic heat alone and on radiogenic heat supplemented by accretionary heat show that its history should be intermediate between those predicted for Rhea (fig. 5.20) and Mimas (fig. 5.23), and in view of its small size, it ought to be more similar to Mimas than to Rhea. The remaining source of heat to which we can appeal as the powerhouse for the activity on Enceladus is tidal dissipation.

We have already seen that the dissipation of tidal energy is an extremely effective way of keeping the interiors of Io and Europa hot. A quick examination of the orbital resonances between the saturnian satellites suggests considerable potential for such a mechanism in Enceladus, because the orbital period of Enceladus is exactly half that of Dione. This means that Enceladus experiences a gravitational "tug" from Dione twice in every orbit. Enceladus's tidal heat flow may be as great as 5 mW m^{-2}. This is three orders of magnitude less than Io's heat flow, but would be enough to initiate partial melting of an ammonia–water mixture at a depth of somewhere between 25 and 100 km below the surface. Furthermore, the rate of tidal heating may vary because of changes in orbital resonances over time within Saturn's large and complicated family of satellites.

Gravitational interactions among more than two bodies are notoriously difficult to calculate, but the most likely multibody scenario can be imagined by considering Janus and Epimetheus. These are a pair of small low-density irregular-shaped satellites each about 150–200 km across, whose orbits lie inside that of Mimas (they are shown but not named in fig. 1.2). Their orbital periods are only fractionally more than half that of Enceladus and are apparently changing as a result of gravitational interaction with Saturn's rings. If either of these were to have been temporarily locked in a 2:1 orbital resonance with Enceladus, while Enceladus remained in a 2:1 orbital resonance with Dione, the forced eccentricity of Enceladus's orbit should have increased sufficiently to enhance considerably the rate of tidal heating. It is possible that Enceladus has experienced several periods of forced orbital eccentricity that lasted, say, between ten and a hundred million years each time, and that each of these was marked by resurfacing of part of the globe.

Whether or not Enceladus is geologically active now, the facts that there have been several episodes of activity with the youngest occurring in the geologically recent past (as attested to by the lack of craters on the sp$_2$ unit), and that the only effective heating mechanism we can think of is probably episodic, make it a pretty good bet that Enceladus will "switch on" again at some time in the geologically near future, within a few hundred million years, if not before.

If the grounds by which Enceladus wins its place in the "active worlds" category are debatable, there can at least be no argument about the final example, Triton, which is the second moon on which eruptive activity has been seen actually in progress.

7.4 TRITON

With Triton we reach the most distant world visited by any space probe from Earth. It was a fitting climax to the *Voyager* project that Triton turned out to be such a fascinating place, with a geologically young and active surface (plate 7). During one week at the end of August 1989, as *Voyager-2* skimmed 5000 km above Neptune's cloud tops and then onward to within 25,000 km of Triton, the attention of the world's media was focused on planetary exploration with an enthusiasm not seen since the *Apollo* Moon landings, and not repeated until the 1997 Mars Pathfinder landing.

Lying on the outer fringes of the solar system, Triton is an icy world the like of which we have not encountered before. It is large and dense (table 1.1), beaten in size only by Titan and Jupiter's galilean satellites, and is remarkable in being the only major satellite of any planet to have a retrograde orbit. That is to say, its orbital motion is in the opposite direction to its planet's spin, and in this case it is also inclined at 21° to the planet's equator. As discussed in chapter 2, Triton's retrograde orbit makes it virtually impossible for it to have accreted from a protosatellite disk around Neptune, so it has long been thought likely that Triton must have formed elsewhere. It is probably a Kuiper belt object that wandered close to Neptune, where collision with another satellite or frictional drag within any remaining gas cloud about the planet slowed its motion sufficiently to allow it to be captured into orbit. Initially this orbit would have been highly eccentric, but tidal interaction with Neptune would eventually have forced it into its present circular pattern. This process may have started immediately upon capture, or there may have been a billion year delay, but either way it probably took on the order of a further billion years to run to completion. While the orbit was being changed, dissipation of tidal energy must have acted as a major heat source (possibly exceeding radiogenic heating by a factor of a thousand or more), leading to global melting and differentiation even within the rocky fraction, so it is likely that Triton now has an inner iron-rich core surrounded by an outer core of silicates.

7.4.1 The Composition of Triton

Triton's faintness and proximity to Neptune make it difficult to study from Earth. Prior to *Voyager-2*'s encounter, methane had been identified in Triton's spectrum, although it was not clear whether it took the form of ice or gas. There was no clear trace of the water-ice absorption features that are so prominent elsewhere, and one faint spectral feature inspired the hope that parts of Triton were covered by oceans of liquid nitrogen. These would have made a fascinating sight, but the reality when *Voyager-2* got there was almost as exciting. The surface temperature turned out to be about 38 K, a little colder than expected (indeed, colder than any other closely observed surface in the solar system) and sufficiently low to freeze nitrogen solid. The presence of scarps up to 1 km in height suggests that the surface has the strength of water-ice, but spectroscopic studies have not revealed any water-ice at the surface, which appears to be covered mostly by nitrogen-ice, giving it a generally pinkish appearance, with up to about a tenth of the surface area covered by carbon dioxide. In contrast, contamination by methane and carbon monoxide appears to be less than about a tenth of a percent, and these species probably occur mostly as solid solutions within the nitrogen-ice. There is a polar cap, apparently of pure nitrogen-ice, that may be a largely seasonal phenomenon.

The whole of Triton's surface has a high albedo, with a global average of 0.78, which is exceeded only by that of Enceladus. Even the darkest material on Triton is really quite reflective, with an albedo around 0.62, whereas the brightest parts of the polar cap have an albedo of 0.89. Triton has a very thin atmosphere, mostly nitrogen and a minor amount of methane. *Voyager-2* determined the atmospheric pressure at the surface to be about 14 µbar, not much more than a hundred thousandth of Earth's atmospheric pressure, but measurements with the Hubble Space Telescope of the dimming of starlight as Triton's atmosphere passed in front of a distant star in 1997 indicate that by then the surface pressure had at least doubled. Triton's atmosphere contains rare thin clouds at altitudes of a few kilometers that were seen mostly over the polar cap and are interpreted as condensed particles of nitrogen-ice (analogous to terrestrial cirrus clouds, which are of water-ice crystals). There is also a pervasive, although tenuous, haze extending to a height of 30 km that is probably a photochemical smog consisting of ethane, acetylene, and the like, resulting from the action of solar ultraviolet ra-

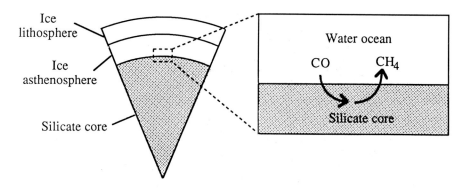

Figure 7.38. Model for the present internal structure of Triton. Radiogenic heat from Triton's largely rocky core is probably sufficient for the melting point of water-ice to lie near the core–mantle boundary, so the lithosphere–asthenosphere transition is likely to be no deeper than the middle of the mantle. The presence of ammonia and/or methane as partial melts may decrease the thickness of the lithosphere still further. The detail shown on the right indicates the situation during tidal heating, when the water-dominated mantle was largely molten and the circulation of hot aqueous fluids through the outer part of the silicate core could have led to the chemical destruction of carbon monoxide (CO), possibly by reducing it to methane (CH_4).

diation on methane and possibly hydrogen cyanide (HCN) from reactions involving nitrogen and methane.

We shall examine some details of Triton's surface shortly, but first it is worth considering its bulk composition and internal structure (fig. 7.38). The *Voyager-2* encounter showed that Triton's density exceeds that of Ganymede and Callisto (table 1.1). Bearing in mind that its lower mass must result in less internal compression, such a high density can be explained only by Triton having a considerably greater ratio of rock (or rock plus metal) to ice than these. In fact, silicates (plus any iron-rich fraction) must make up between about 65 and 75 percent of the total mass, greater than for any other major outer planet satellite except Io and Europa.

In Triton's case, it is simplest to regard this ratio as reflecting a decrease in the amount of water available to condense from the solar nebula rather than an excessive abundance of silicates. The latter would be hard to explain, but the low abundance of water can be accounted for by arguments based on chemical reaction rates, which suggest that in the cold outer region of the solar nebula where Triton probably formed (and in the absence of local additional heating from the condensation of a neighboring giant planet), much of the hydrogen and oxygen would have been partitioned into the phases H_2 and CO (hydrogen and carbon monoxide) in preference to H_2O and CH_4 (water and methane). With relatively little water available to condense while Triton was forming, the rock-to-ice ratio became higher than in icy bodies that developed closer to the Sun. The same situation would favor N_2 (nitrogen) over NH_3 (ammonia), which accords with Triton's observed surface composition.

One problem with this model is that under these conditions we would expect Triton to contain about as much carbon monoxide as nitrogen, whereas in fact its carbon monoxide-to-nitrogen ratio appears to be extremely low. The explanation probably lies in chemical reactions that would have occurred between Triton's ice and rock fractions during the global heating event. This could have sustained the existence of a mantle of liquid water for a considerable period, although it is likely to be almost entirely frozen by now. While the "ocean" existed, tidal and radiogenic heat in the core would have been carried away convectively by the drawing down of cold water through fractures in the rock, whereupon it would have been heated to escape back into the ocean through hydrothermal vents. While this was going on, carbon monoxide that was in solution would

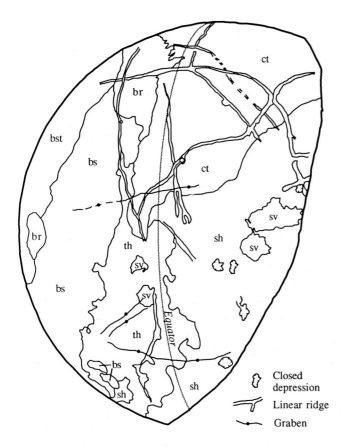

Figure 7.39. Geological sketch map of terrains on Triton (same area as plate 7). *sh*, high smooth plains; *sv*, smooth floor material; *th*, hummocky plains; *ct*, cantaloupe terrain. *bs*, *br*, and *bst* are divisions of the polar cap (bright spotted, bright rugged, and bright streaked). See sections 7.4.2 and 7.4.3 for explanation.

have had the opportunity to react with the warm, hydrated rock. Chemical models show that, irrespective of whether conditions were oxidizing or reducing, carbon monoxide is likely to have been destroyed. Under oxidizing conditions it would be converted to carbon dioxide, whereas under reducing conditions it would be converted to simple organic molecules or possibly methane. Interestingly, methane is the most abundant gas dissolved in the water escaping from hydrothermal vents such as black smokers at active spreading axes in Earth's oceans today, but most of this is primordial methane that has been trapped within the Earth since its accretion, and little, if any, is the immediate product of the reduction of carbon monoxide as proposed for Triton.

7.4.2 The Surface of Triton

Because of the high inclination of Triton's orbital plane, the northern polar region was in shadow in 1989. *Voyager-2* imaged as much of Triton's Neptune-facing hemisphere as was then sunlit with a best resolution of about 400 m per pixel (plate 7). The low latitudes are occupied by several distinct terrain types, which are overlain by a south polar cap that is thin enough near its edges to allow the morphology of the underlying terrain to show through.

The distribution of Triton's main terrains is shown in figure 7.39. The three most widespread units are the high smooth plains (sh), hummocky plains (th), and the rather curious cantaloupe terrain (ct). It is fairly clear that the high smooth plains unit overlies the other two, and in places the images show lobate scarps a few hundred meters to a couple of kilometers high, both at its margins and within it. This terrain has every appearance of having been emplaced as a series of large-volume, highly viscous cryovolcanic flows, perhaps of one of the kinds suggested for Ariel in section 6.4.3. With proven nitrogen, carbon monoxide, carbon dioxide, and methane, and presumably also ammonia, available to form intergranular fluids and a variety of watery partial melts, there is scope for an especially wide range of volcanic phenomena during Triton's history. Several roughly circular depressions and chains of both rimmed and rimless pits are the probable sources of these outpourings.

A similar but low-lying unit is emplaced within both the high smooth plains and the hummocky plains, occurring in the floors of four flat-floored depressions up to about 200 km across. This has become known as smooth floor material (sv). The depressions are bounded by embayed escarpments, reminiscent of the escarpments in part of the plains terrain of Io (section 7.1.1). In places the escarpments are terraced where one segment of wall overlaps another (fig. 7.40). It is not clear how much the present expression of the edges of these depressions is the result of constructional volcanic processes, how much was caused by caldera collapse, and how much is a product of the removal of overlying layers by some kind of mass-wasting phenomenon, perhaps related to the loss of volatiles as may occur on Io (section 7.1.1). In the center of each depression is a rugged area near the limit of the image resolution, which appears to consist of a collection of pits and flows and to be the site of the most recent volcanic activity.

The hummocky plains unit emerges from between the high smooth plains and the polar cap in Triton's leading hemisphere. It appears to be another flow-generated terrain containing a variety of domes, and a 700 km long raised ridge named Vimur Sulcus that looks as if it was constructed by viscous flows at an eruptive fissure. A detailed view of hummocky terrain is shown in figure 7.41. The hummocky plains exhibit the greatest concentration of obvious impact craters on Triton, with a crater density comparable to that seen on the lunar maria. The crater density decreases away from the apex of orbital motion. Unfortunately, this tells us nothing of the origin of the impactors, because Triton's retrograde orbit means that its leading hemisphere must take the full brunt of impacts whether from debris orbiting Neptune or from material external to the Neptune system. It would be unwise to use comparisons of Triton's crater density with that of the lunar maria to imply an absolute age for the hummocky plains, but this terrain appears to be at least twice as old as Triton's much more lightly cratered high smooth plains.

The other major terrain is of a type unlike any seen elsewhere in the solar system (fig. 7.42). This consists of a dense patchwork of slightly elliptical dimples, mostly 20–30 km across, traversed by a network of ridges that appear to be viscous extrusions erupted along fault systems. Because of the similarity in appearance to the skin of a well-known variety of melon, this has become known as the cantaloupe terrain. The origin of the dimples is a mystery. It is unlikely that they are impact craters mantled or otherwise modified by later events, because they are all too similar in size to one another. More credible suggestions are that they are a result of melting and collapse of the icy surface, or explosive craters like terres-

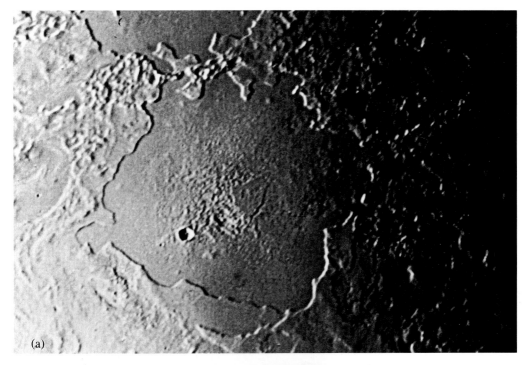

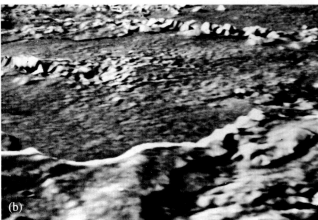

Figure 7.40. (a) A smooth plains depression within Triton's high smooth plains unit named Ruach Planitia. Note the overlapping terracing at the bottom of the image. The crater toward the lower left of the depression is 15 km across and is probably an impact structure, whereas the finely textured area to its right may be evidence of volcanism. (b) A synthetic perspective view looking diagonally across Ruach Planitia, seen from the lower right of (a).

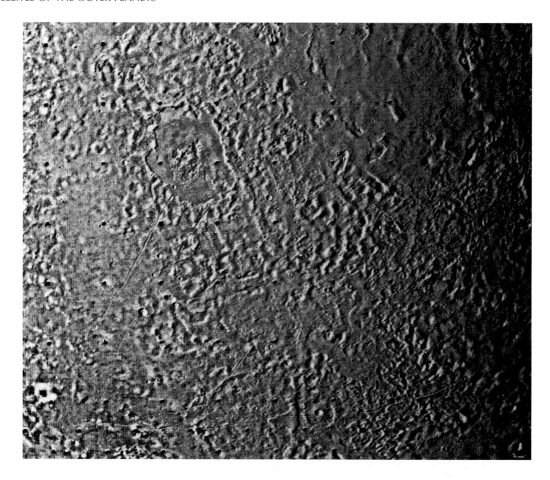

Figure 7.41. Part of Triton's hummocky terrain, from the central lower parts of figure 7.39 and plate 7. The ridge Vimur Sulcus, apparently a volcanic fissure, extends upward from the lower edge of the view, skirting a smooth-floored depression (Sipapu Planitia) above left of center. Two nearly straight and parallel narrow troughs (mapped as grabens in fig. 7.39) named Jumna Fossae run into Sipapu Planitia from the lower left, and similar features can be seen elsewhere in the lower part of the image. High smooth plains occupy the upper right portion of the image.

trial maars (where magma causes groundwater to explode as steam), or the surface expression of pods of less dense ice (perhaps pure H_2O) that have risen as diapirs to pierce through near-surface layers of ices made denser by the presence of carbon dioxide or ammonia. The evidence of superposition suggests that the cantaloupe terrain is the oldest unit on Triton and has been partly buried by both the smooth and the hummocky plains units. Even allowing for the fact that it lies in the less cratered trailing hemisphere, it is curious that no definite impact craters have been identified within it.

To judge from the pattern of linear ridges in the cantaloupe terrain and elsewhere, Triton's tectonics appear to be dominated by extension. These features

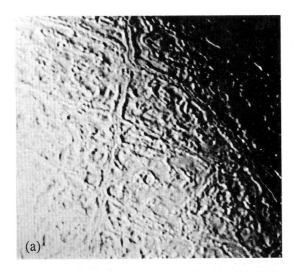

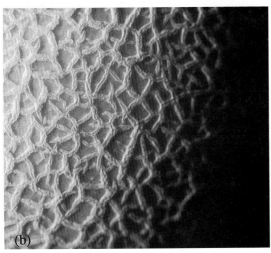

Figure 7.42. (*a*) View of part of Triton's mysterious cantaloupe terrain, from the upper right of plate 7. Several overlapping ridges may be made out, superimposed on a dense array of near-circular dimples. The image is about 450 km across. (*b*) Not part of Triton, but a close-up of the skin of a cantaloupe melon. The similarity in appearance is striking, although it casts little light on how Triton's cantaloupe terrain formed. Indeed, the comparison is doubly misleading, because the presence of methane and ammonia would almost certainly impart a greatly inferior flavor to Triton, compared to that of a real melon.

seem to be fissures that became flooded and indeed overfilled by viscous extrusions. In addition, there are a few narrow troughs on the order of a kilometer across and up to 1000 km long that cut all units and are probably the most recent faults. These cause no offsets and are mapped as grabens on figure 7.39. Some workers have suggested the action of strike-slip tectonics near the boundary between the cantaloupe terrain and the smooth plains, but no evidence of a surface fault-break is visible at the resolution of the available images.

In summary, then, the terrain units on Triton are, in the plains areas at least, dominated by the extrusion of viscous, but often fairly extensive, cryovolcanic flows. On the whole, the surface appears to be relatively young, but despite the fascinating range of landforms, there is no evidence of current activity. For this, we have to examine the polar cap.

7.4.3 The Polar Cap

Both Earth and Mars have polar caps, and the extent of these changes with the seasons. Frost migration on Callisto and Ganymede notwithstanding, the only known seasonal polar caps in the outer solar system appear to be on Triton, where the probable seasonal ice is nitrogen instead of water-ice (as on Earth) or water-ice and carbon dioxide (as on Mars). The *Voyager* images showed much of Triton's southern hemisphere to be occupied by a bright polar cap, with an albedo of around 0.88 and a ragged edge suggesting that it may have been retreating at the time. To understand this it is necessary to appreciate that Neptune's (and Triton's) year is 165 Earth-years long and that a combination of Neptune's tilt and the inclination of Triton's orbit means that the subsolar latitude on Triton varies from about 55° N to about 55° S during the year. Such a large range leads to the curious situation that Triton's tropics (55° N and S) lie farther from the equator than its "arctic" and "antarctic" circles (35° N and S). At the time of the *Voyager-2* flyby, Triton's southern hemisphere was about three-quarters of the way through its 40-year spring, and seasonal ice that had accumulated during the long darkness of the winter was being burned off by the Sun to be redeposited in the northern hemisphere.

This process offers a way to explain changes in Triton's spectrum. In 1979 it showed clear signs of methane, but during the following decade the methane absorption features became obscured and the spectrum began to show those traces of nitrogen that inspired the short-lived hopes of nitrogen oceans. In the light of the *Voyager* data, at least part of this spectral change can be attributed to the result of the migration of nitrogen from the southern polar cap and its redeposition in the northern hemisphere, which in 1979 still had regions of exposed methane deposits that had lost their nitrogen cover during the previous northern summer. Unfortunately, the change that this simple process could make in Triton's spectrum appears to be less than what was actually observed, and it is necessary to call for the deposition of fresh nitrogen frost over at least the fringes of the bright polar cap as well. This last point lends credence to models suggesting that the bright polar cap is actually rather stable in the long term, because it is formed of ices whose crystalline structure makes them highly reflective and therefore prevents the surface from heating sufficiently for the full thickness of the cap to be completely lost by volatilization before the end of the summer. Seasonal changes, according to these models, are manifested mainly by the migration of a darker nitrogen frost that overlies the bright polar cap in winter and is volatilized and redeposited in the opposite hemisphere during the spring and summer.

Thus, Triton's bright southern polar cap may not be the simple seasonal affair that it appears to be at first sight, in which case the seasonal changes are probably of considerable complexity. It is a pity that too little of the northern hemisphere was in sunlight at the time of the *Voyager* encounter to show whether or not there was a northern polar cap as well. In the event our observational evidence for polar processes on Triton comes from the southern hemisphere alone. In figure 7.39 three units are distinguished within the southern polar cap: bright rugged (br), bright spotted (bs), and bright streaked (bst). The bright rugged unit appears simply to be cantaloupe terrain with bright polar material resting in the hollows. If the bright polar cap shrinks as Triton's southern spring progresses

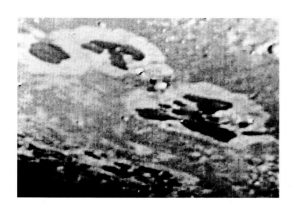

Figure 7.43. Possible isolated areas of Triton's bright spotted unit (from the lower part of plate 7), apparently left behind by the retreating polar cap. Alternatively, these could be dark viscous flows surrounded by bright aureoles. The mushroom-shaped dark patch in the foreground (Akupara Maculae) is about 100 km across.

into summer, it is likely that an increasing extent of cantaloupe terrain will be uncovered. Apparent evidence that the bright polar cap is in either long-term or seasonal retreat is provided by the bright spotted unit, where there is incomplete cover by polar ices. Here, roundish patches of the underlying surface show through, apparently on high ground, with thicker opaque ices in the low areas. This underlying unit is darker than most of what is exposed to the north, beyond the limits of the ice cap. There is one curious area (fig. 7.43) where dark patches surrounded by pale aureoles appear to be outliers of the bright spotted unit isolated by the retreat of the polar cap, though there are alternative explanations for this pattern, such as lobate, viscous flows.

The bright rugged and bright spotted units pass poleward into the bright streaked unit, which appears to be thick enough to obscure the underlying terrain completely. This unit, and parts of the spotted unit, is marked by streaks of dark material ranging from a few tens to about a hundred kilometers in length. Several of these can be seen on plate 7, mostly oriented toward the northeast, except within 10° of the pole, where they point in all directions and may cross one another. Typically, a streak fans out slightly down-length, and usually has a dark patch a few kilometers across at its head. The dark color of the streaks and dark spots may be caused by complex organic polymers, such as tholins, produced from methane by the action of charged particles trapped in Neptune's magnetic field, or by cosmic rays and ultraviolet photons.

If they lie on top of the seasonal ice, these streaks must be ephemeral features, but if the bright polar cap is stable, then many may date from previous years. They look similar to streaks formed by deposits of windblown dust on Mars, and it has been calculated that on Triton grains 5 μm in diameter or less could be carried in suspension by winds of about 10 m s^{-1}, the thinness of the atmosphere being compensated for by Triton's low surface gravity (0.78 m s^{-2}, less than a tenth of Earth's gravity). Such a speed is force 5 (a fresh breeze) on the terrestrial Beaufort scale and is well within the bounds of possibility on Triton.

Without attempting a prolonged discussion of Triton's meteorology, it is instructive to make a few comparisons with other planets. On Mars the atmosphere contains about three to five times as much mass as the seasonal polar caps, but on Triton the atmosphere has only about a hundredth of the mass of its observed seasonal cap (depending on how much of the cap is actually seasonal).

This means that Triton's atmosphere can be regarded merely as the polar material in transit: whatever is sublimed from the south polar cap as it shrinks must be transported northward to be deposited in the growing north polar cap at a roughly equivalent rate. If a major fraction of the south polar cap is to be removed before the end of the summer, then the time spent by each nitrogen molecule in transit can be no more than a small fraction of a Triton year. Consequently, the atmospheric pressure is likely to be very unstable, and it would be perfectly feasible for it to increase or decrease by a factor of ten or more from its value at the time of the *Voyager* encounter.

As the south polar cap sublimes the local atmospheric pressure must increase, and winds will develop that cause a net transport of vapor away from the region (the doubling of global atmospheric pressure between 1989 and 1997 suggests that the mass of the atmosphere more or less doubled over the same period). However, on a rotating body such as Triton such winds cannot simply blow directly from south to north. Instead, there is likely to be (at least near ground level) a polar anticyclone with winds blowing northeastward at about 10 m s^{-1} in a spiral pattern round the pole. This deflection of the winds to the right in the southern hemisphere is the opposite of what would be found on any other body in the solar system, because, as a result of its retrograde orbit and captured rotation, Triton rotates from east to west instead of from west to east.

Thus, the generally northeast orientation of the dark streaks on the south polar cap is compatible with anticipated wind directions, and the streaks themselves can be explained if the black patches usually found near their heads are sites where suitable particles can be picked up. Although the origin and composition of this material are uncertain, it is clear that if these particles are picked up from the ground by the wind, then they must display a remarkably low degree of cohesiveness. On Earth fine-grained particles tend to stick together because of a thin film of water on their surface and other molecular forces, but if such cohesion were to occur between particles on Triton it would be impossible for the anticipated 10 m s^{-1} winds to pick them up.

7.4.4 Plumes

One way to sidestep the problem of surface cohesion of fine-grained particles is to have the particles thrown up into the air by some kind of eruptive process, because this would probably separate even cohesive grains, which, once airborne, would be at the mercy of the winds. Two such eruptions on the polar cap were imaged during the *Voyager-2* flyby, one of which is shown in figure 7.44. Each of these was powerful enough to send a jet of dark material vertically upward to an altitude of 8 km, at which height it bent sharply over to become a dark horizontal cloud extending westward for over 100 km. Like certain volcanic eruption plumes on Earth, it appears that Triton's plumes cease to rise when they reach an altitude at which they are no longer buoyant; probably in this case the 8 km limit marks the tropopause, above which the temperature of the atmosphere starts to increase.

The fast westward-blowing winds at high altitudes required to account for the orientation and length of the horizontal part of the plumes can actually be made to tally with models calling for slower winds blowing northeastward at ground

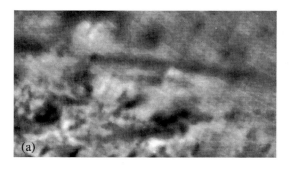

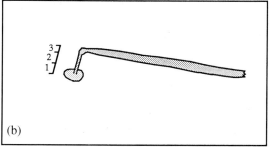

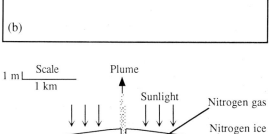

Figure 7.44. (a) *Voyager-2* image of a dark plume erupting over Triton's south polar cap. Its source is a dark spot on the surface that has been named Mahilani, and it rises to 8 km before dispersing downwind at a constant altitude. (b) Sketch interpretation of the plume shown in (a). In the height interval marked 1, the plume is rising through a combination of its erupted momentum and buoyancy forces; in the height interval marked 2 it is rising mainly through buoyancy, because it is warmer than the surrounding atmosphere. At the tropopause (height 3) the atmospheric temperature gradient reverses and the plume becomes neutrally buoyant, so it ceases to rise. The wind at this altitude then streaks the plume out over a length of some 100 km.

Figure 7.45. A possible eruption mechanism for Triton's plumes, here interpreted as a geyserlike eruption driven by the pressure of nitrogen gas escaping from beneath a "greenhouse" layer of nitrogen-ice. Note the vertical exaggeration.

level. However, particles falling out from such a plume would produce an elongated dark deposit over the surface ice with an orientation quite different from that commonly shown by the dark streaks. Possible explanations are that most of the dark streaks come from dust picked up directly from the surface by the wind and that only a few result from plumes of the type seen in the images, or (not unreasonably) that there were changes in Triton's weather pattern earlier in the spring before *Voyager-2* arrived.

As to the cause of the plumes, there are probably as many explanations as there are scientists who have studied them. They could originate as dust devils driven by solar heating of dark surface patches, but their source at the polar cap argues strongly that nitrogen is the driving gas, although methane remains a possibility. The energy source could be from within Triton, either in the form of the background heat-flow or from the intrusion of comparatively warm icy lavas below the polar cap, either of which could vaporize its base to liberate nitrogen gas. Such mechanisms would be closely analogous to geysers on Earth, where volcanically enhanced heat flow causes the flash vaporization of water to steam. An alternative, suggested by the position of the two observed plumes close to the latitude where the Sun was overhead at midday, is that solar energy is responsible. This model is illustrated in figure 7.45, which shows a layer of nitrogen-ice about 2 m thick overlying a dark substrate. The ice is transparent and allows most of the sunlight to penetrate to the dark layer, where it is absorbed and converted to heat. A rise of only 10 K above the surface temperature of 37 K would cause the vapor pressure of nitrogen to rise about a hundredfold, so the conduction of heat into the ice from the solar-heated dark substrate could allow a large volume of pressurized nitrogen vapor to build up, probably in interconnecting fissures rather than in a vast blister as implied in the simple sketch in the figure. Eventually, the solid nitrogen cap would rupture, allowing gas to escape rapidly, accelerating as it decompressed and picking up dark particles and other material from below and within the exit nozzle.

It would certainly be fascinating to return to Triton later in the southern hemisphere summer, to see what remains of the southern polar cap, whether any of the dark streaks have survived the removal of the ice, and whether any plumes are active. For the foreseeable future, however, we shall have to make do with what *Voyager-2* showed us. Beforehand, most of us would probably have settled for one look, but now it is clear that Triton is one of the most fascinating worlds in the entire solar system, which many would now place high on their repeat wish list.

8 Unseen Worlds

What is this world? what asketh men to have?

Chaucer,
The Knightes Tale

We have now discussed all the major satellites in the outer solar system of which the existing images permit us make direct geological inferences. There remain just three more bodies to consider in our survey of the geology of outer planet satellite systems. These are Titan, the largest satellite of Saturn, and the remarkable double system of Pluto and Charon. The former is wreathed in a nearly opaque atmosphere and the latter pair has not yet been visited by a space probe, but despite this we are gradually learning more about them. Finally, we will speculate about the multitude of world-sized icy bodies in the trans-neptunian space of the Kuiper belt and beyond.

8.1 TITAN

Titan was discovered in 1655 by the Dutch scientist Christiaan Huygens. Although it was the next satellite to be found after the galilean moons of Jupiter, it has kept many of its secrets up to the present day. Titan's methane-rich atmosphere was discovered in 1944 by Gerard Kuiper (another Dutchman, although working in America) by means of spectroscopic techniques, and several other constituents were identified prior to the *Voyager* encounters.

Titan is a large world, in the same class as Ganymede and Callisto in terms of both size and density (table 1.1). The radius of its solid surface is about 2575 km, only 60 km less than that of Ganymede. As its atmosphere is opaque in the visible region of the spectrum to an altitude of some 200 km, this thickness is sometimes (rather unreasonably) added to the solid radius to give Titan the honor of being the largest satellite in the solar system.

As the only solid body beyond Mars with a substantial atmosphere, and an

197

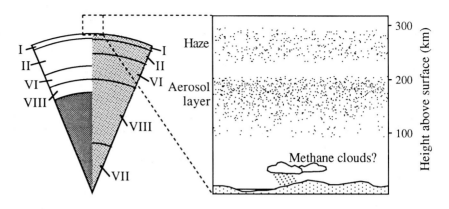

Figure 8.1. Alternative models for the internal structure of Titan, assuming a 52:48 (by mass) rock-to-ice ratio, showing the phases of water-ice that would be stable. On the left is a differentiated model, with all the silicates segregated into the core; on the right, an undifferentiated model, which is less likely. There may also be an iron-rich inner core. The box shows the structure of the atmosphere.

organic-rich one at that, Titan was a target of prime importance for *Voyager* investigation. *Voyager-1* was sent on a trajectory that took it through Titan's Earth and Sun occultation zones, the former so that radio signals from the probe could pass through Titan's atmosphere en route to Earth, and the latter so that the atmosphere could be studied by looking through it toward the Sun. These constraints made it necessary for the probe to continue over Saturn's south polar region, where the gravitational slingshot effect flung it northward out of the plane of the ecliptic, precluding any further planetary encounters.

During the hours before closest approach, *Voyager-1* imaged most of Titan's illuminated disk with a resolution better than 2 km per pixel, and half of it better than 650 m per pixel, but no breaks in the cloud were to be seen. Instead, Titan was shown to exhibit a completely opaque red aerosol-rich layer 200 km above the ground surface, consisting of microscopic droplets of hydrocarbons, with several haze layers above (plate 8). The *Voyager* images show the aerosol layer to be darkest in a hood around the north pole and somewhat brighter in the southern hemisphere than the northern (apparently a seasonal feature, because Hubble Space Telescope images in the 1990s showed the asymmetry reversed), but on the whole Titan is bland in the extreme at visible wavelengths, with just a few east–west markings that show up under extreme contrast enhancement.

Voyager-1's gravitational interaction with Titan gave the first accurate determination of its mass, and the radio occultation data measured its radius to within an error of about 2 km. These data suggest a rock-to-ice ratio of about 52:48. However, because we lack the crucial evidence of Titan's surface morphology that would have been provided by close-up images of its surface, we are unable to say very much about its geological history. Models for its present structure are shown in figure 8.1; accretional and radiogenic heat were probably sufficient to allow Titan to differentiate to some extent at least, especially if the ice was originally volatile-rich as implied by the thickness and composition of the atmosphere.

Both *Voyagers* probed the nature of Titan's atmosphere with a suite of instruments spanning the spectrum from the ultraviolet to radio wavelengths. The attenuation of the radio signal from *Voyager-1* as it passed behind Titan suggests that the surface temperature and atmospheric pressure are 94 K and 1.5 bar, respectively. The latter value means that Titan's ground-level atmospheric pressure is 50 percent greater than at sea level on Earth. Ultraviolet and infrared spectrometer data indicate that Titan's atmosphere consists mostly of nitrogen with between 2 and 10 percent methane. Other species that have been detected are hydrogen (about 0.2 percent) and smaller traces of other, mostly organic gases, including ethane (C_2H_6), propane (C_3H_8), acetylene (C_2H_2), hydrogen cyanide (HCN), and carbon monoxide (CO).

Argon has not been detected, but it is accepted that argon is required in the atmosphere at about the one percent level to increase the average molecular weight of the mixture to the value implied by the radio occultation data. However, it is unlikely that another of the inert gases, neon, is present in any amount, because this would decrease the average molecular weight to below what is compatible with the data. This has important implications for the origin of Titan's atmosphere, which ought to have roughly equal abundances of nitrogen and neon had it been captured directly from the solar nebula or the proto-Saturn nebula. The lack of any large fraction of neon shows that the atmosphere almost certainly formed by degassing out of the ice and rock of which Titan is composed.

The original ice would have been a clathrate, with methane, nitrogen (perhaps in the form of ammonia), and argon (if present) held in voids within the water-ice lattice. After degassing, which was probably associated with internal differentiation, solar ultraviolet photons would have begun to break down any ammonia to nitrogen (N_2), which is heavy and has remained in the atmosphere today, and hydrogen (H_2), which is light and has almost all escaped. In the meantime, if ammonia ever had the chance to build up to a significant concentration in the atmosphere, it would probably have acted as a "greenhouse" gas, trapping solar heat and raising the surface temperature by tens or even hundreds of degrees. Apart from breaking down the ammonia, ultraviolet light has probably always played an important role in modifying Titan's atmosphere, in driving photochemical reactions by removing hydrogen atoms from methane and thus initiating the assembly of the larger organic molecules and cyanide compounds found today.

One of the most exciting aspects of Titan's atmosphere is that conditions are too cold to allow a significant amount of water vapor to exist. Consequently, methane has not had the opportunity to react with water vapor and be converted to carbon dioxide, which would have turned the atmosphere into a carbon dioxide and nitrogen mixture such as we find today on Venus and Mars and which is inferred for the prebiotic Earth. Titan thus offers a unique opportunity to study a planetary atmosphere that may have begun in a state similar to that of the inner planets but whose evolution followed an entirely different route.

Although we have no direct information on what goes on below the aerosol layer, the weird conditions of Titan's atmosphere ought to allow both methane and ethane to condense. In particular, it has been calculated that, over time, enough ethane should have been made in the upper atmosphere to have drizzled downward and formed a liquid layer about 1 km deep. Apart from the background drizzle of ethane originating at high levels, there may also be clouds of methane condensing lower down, giving rise to either methane snow or methane rain. So, if we could see Titan's surface, we might well find it to be sculpted by the action of winds, flowing liquids, and wave action, resulting in an enormous range of erosional and depositional processes, similar to those familiar on Earth.

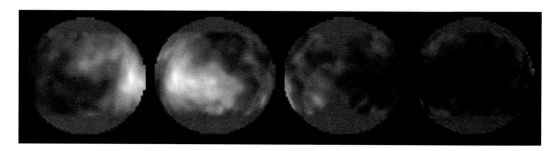

Figure 8.2. Four hemispheric views of Titan's surface as seen by the Hubble Space Telescope WideField/Planetary Camera in the near-infrared taking advantage of low-opacity "windows" in Titan's atmospheric haze near 0.94 and 1.08 μm. From left to right: anti-Saturn hemisphere, leading hemisphere, Saturn-facing hemisphere, trailing hemisphere. These views were made by combining fourteen separate images, and the effective resolution is approximately 500 km. The polar regions could not be mapped because of the viewing angle and the projected thickness of the haze. The large bright area in the leading hemisphere is about as big as Australia. The *Voyager-1* radio occultation experiment showed that the difference in height between this bright area and the dark area to its west must be less than about 1 km.

Such considerations raised hopes of finding another world with oceans, but these were dashed in June 1989 when, using Deep Space Network antennas that were needed immediately afterward for communications with *Voyager-2* during its approach to Neptune, a radar beam was bounced off Titan's surface. The strength of the returned signal was stronger than expected from a global surface ocean even as shallow as 200 m, and was more like an echo from a dry, rough surface. Intriguingly, changes in polarization of the backscattered radar signal from Titan are more like what would be expected from an ammonia–water ice than from purer water-ice. Global images of Titan's surface were obtained in 1994, using the Hubble Space Telescope to record near-infrared light that is able to penetrate Titan's atmospheric haze (fig. 8.2). These images showed an apparently stable albedo pattern (varying from place to place by about 10 percent) indicative of a largely solid surface, although small seas and lakes could still be there in abundance.

A continual fallout of tholins and other organic molecules is expected to accompany the ethane and methane precipitation from Titan's atmosphere, and this should be sufficient to make the surface uniformly dark over a timescale of less than a hundred thousand years. Thus, the bright areas are probably recently resurfaced, perhaps being washed clean of dark organic deposits by methane rainfall, which in equatorial regions could increase by an order of magnitude for a 500 m altitude difference. Alternatively, cryovolcanism could be responsible, though an exceptionally large and rare eruption would be required to explain the largest bright area in figure 8.2.

The prospects of cryovolcanism on Titan are intriguing. There is no significant source of tidal heating, but radiogenic heat could be sufficient to stimulate partial melting of Titan's icy mantle, which should be more volatile-rich than that of Ganymede or Callisto. Titan's atmospheric pressure is so high that volatiles such as methane could not escape explosively from any cryomagmas reaching the surface. Pyroclastic deposits are therefore not expected; instead, cryovolcanic landforms should consist of viscous flow features. Despite the inhibition of explosive eruptions, cryovolcanism would be an effective way of transferring methane from Titan's interior to its atmosphere, and it can be argued that the rate of photochemical conversion of methane to ethane is such that the present atmospheric concentration of methane can be explained only by replenishment by outgassing from the interior. As described in chapter 9, images from the *Cassini/Huygens* mission should give us our first good view of Titan's geology and surface processes in 2004.

8.2 PLUTO AND CHARON

For the first forty years or so after its discovery in 1930, Pluto, the ninth planet, was regarded as a solar system misfit: too small to be a gas giant and too far from the Sun to be a terrestrial planet—interesting, no doubt, but no place for a geologist. Now it seems to fit neatly into place as the largest of the Kuiper belt objects, and there are strong grounds for expecting its surface to have been sculpted by a complex array of processes.

Pluto appears so small in a conventional ground-based telescope that the diameter of its disk cannot be measured with any accuracy, and until the mid-1950s the best estimates were that it was about half the size of Earth, or perhaps slightly bigger. This, then, could hardly be the ten-Earth masses body required to cause the perturbations in Neptune's orbit that led to the tracking down of Pluto in the first place, which now appears to have been a piece of good luck as much as anything.

The accepted size of Pluto began a further dive when, in 1976, solid methane was detected in Pluto's spectrum. It became clear that the surface is icy and comparatively highly reflective, calling for a small high-albedo body rather than a larger low-albedo body. Two years later, high-resolution telescopic photographs revealed the presence of a satellite in orbit around Pluto. Given the name Charon, which it shares with the ferryman of classical mythology who conveyed the dead into the god Pluto's underworld domain, this satellite soon began to provide important information. The dimensions and period of its orbit enabled the combined mass of the Pluto–Charon system to be calculated, showing that Pluto must be far less massive than Earth, and indeed is outdone even by Triton.

How Pluto and Charon came to be together as a pair was at first mysterious. Separate origins followed by gravitational capture would require a seemingly too improbable sequence of events, and co-accretion as a double planet is unable to account for the Charon's large mass and orbital angular momentum. It is now accepted that Charon probably accreted from the debris resulting from a giant impact between Pluto and a smaller Kuiper belt object with a mass about a tenth to half of Pluto's mass. This debris would have consisted of the fragmented remains of the impacting object and parts of Pluto's outer layers. Some of this debris would have fallen back onto Pluto, but much of it would have clumped together in orbit around Pluto to form Charon. Earth's Moon is believed to have formed in a similar way, following a collision between the proto-Earth and a large planetesimal.

Charon's orbit about Pluto is inclined at 94° to Pluto's orbit about the Sun, so technically it is retrograde. Tidal interactions mean that the plane of Charon's orbit coincides with the plane of Pluto's equator, so the 94° inclination shows that Pluto itself is tipped on its side. Furthermore, Charon's orbital period of 6.4 days is the same as the rotation period of Pluto as deduced from periodic variations in its brightness. Thus Pluto always keeps the same face turned toward Charon, and presumably Charon's rotation is tidally locked to Pluto in a mutual fashion.

Information about Pluto and Charon was boosted during 1985–1990 when on many occasions the line of sight from Earth passed through the plane of Charon's orbit about Pluto, so that Charon passed directly behind and in front of Pluto. By

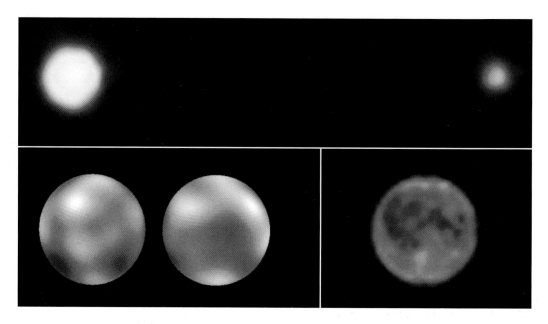

Figure 8.3. (*Top*) Pluto and Charon imaged by the Faint Object Camera (FOC) of the Hubble Space Telescope in February 1994, showing their relative sizes and separation. (*Lower left*) Views of western and eastern hemispheres of Pluto, constructed by enhancing Hubble Space Telescope FOC images obtained in June and July 1994. Pluto's axis was tilted slightly toward Earth, so the north pole is just visible but the south pole is out of sight in both views. (*Lower right*) The Earth-facing hemisphere of the Moon, degraded to a similar resolution.

timing and measuring the changes in combined brightness during these events, the relative sizes of Pluto and Charon could be calculated. This series of occultations was a stroke of good luck, because such a spell of favorable alignments occurs only twice in Pluto's 248-year orbit about the Sun, and if Charon had been discovered just twelve years later we would have missed it. The best images to date of Pluto and Charon have been obtained by the Hubble Space Telescope, and these can be enhanced to show the general distribution of surface albedo patterns (fig. 8.3).

Estimates for the sizes, masses, and density of Pluto and Charon are given in table 1.1. Their relative masses differ by a factor of less than ten, making Charon by far the most massive satellite with respect to its planet. Previously, this honor went to our own Moon. Charon orbits closer to its planet than any other satellite in the outer solar system (large or small), and seen from the surface of Pluto it would have an angular diameter of almost four degrees, eight times the apparent diameter of the Moon in Earth's sky. The development of such a close relationship is likely to have been associated with considerable tidal heating, and it is probable that both bodies have had the opportunity to become fully differentiated. However, if (as is likely) the Pluto–Charon pair owes its origin to a giant impact, then this initial collision would have melted and therefore differentiated Pluto right from the start.

Pluto's surface shows more large-scale contrast than any planet except Earth, and has a bright cap around the better-seen north pole. Methane, nitrogen, and carbon monoxide have all been identified spectroscopically. The bright areas have the albedo of clean snow on Earth and are probably fresh nitrogen frost deposits with limited contamination by methane and carbon monoxide, whereas the dark areas are likely to be tholin-rich residues and are apparently warmer than anywhere on Triton. Charon is bluer than Pluto and has a lower albedo. As

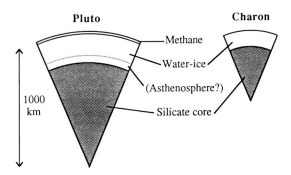

Figure 8.4. Possible internal structures of Pluto and Charon. The best estimate for Pluto's bulk composition is that it has 68–80 percent silicates, with the balance made up of ices. Where water-ice is shown, this could include several percent nitrogen and carbon monoxide. Most of Pluto's methane crust is obscured by nitrogen frost.

far as can be told, its spectrum is lacking in methane features and appears to show water-ice only. If they are differentiated, Pluto and Charon should each have a sizable rocky core overlain by a mantle of water-ice and topped, probably in the case of Pluto only, by methane (fig. 8.4).

The similarities in size, density, and composition between Pluto and Triton are striking. Further similarities emerge when we consider Pluto's atmosphere. The presence of an atmosphere around Pluto was proved when, in June 1988, the passage of Pluto in front of a star was observed for the first time. The light from the star began to fade gradually as it was attenuated by Pluto's atmosphere, suggesting an atmospheric pressure less than a hundred thousandth of Earth's, but similar to Triton's. Apart from nitrogen, the atmosphere may also contain less volatile gases such as methane and carbon monoxide.

Nitrogen is also the major component of the atmosphere of Triton, and like Triton, Pluto experiences seasonal changes so severe that the bright surface patches are most likely to be transient features representing migratory frost deposits. Apart from Pluto's 94° axial tilt, its orbit is highly eccentric, so during its year its distance from the Sun varies from 7.4 billion kilometers (at aphelion) to 4.5 billion kilometers (at perihelion), producing extreme variations in the global mean temperature. Pluto reached perihelion in 1989 (being closer to the Sun than Neptune during the years 1979–1999), and it is probable that its atmosphere is a temporary phenomenon, being volatilized from the surface only during the sixty or so years closest to perihelion.

On the other hand, it has been argued that during the half of its year when Pluto is closest to the Sun, methane (which is lighter than nitrogen) will escape from its atmosphere to space at a significant rate. To maintain a long-term steady state, the atmospheric methane would need to be replenished by a net loss of about a millimeter of methane-ice from the surface per orbit, implying a loss of approaching 10 km of surface methane over the lifetime of the solar system. This has been used as a powerful argument in favor of Pluto being fully differentiated, because if even a tiny fraction of particles of rock or water-ice were mixed with the crustal methane layer these should, over only a few Pluto-years, form a surface lag deposit thick enough to insulate the underlying methane from solar heating. Thus, according to this logic, Pluto appears to have an outer layer of rather pure methane that, if it was formed early in the lifetime of the solar system, must have originally exceeded about 10 km in thickness. This is not far short of the total amount of methane expected in a body of Pluto's mass on cosmochemical grounds. Charon, with its lower gravity, has evidently lost all the methane it ever had available near the surface.

If the rate of loss of methane from the surface of Pluto has just been correctly described, then (apart from migratory frost patterns) Pluto may be a rather featureless world. Primordial craters less than about 5–10 km deep will have been erased, together with any ancient topographic features resulting from tectonic and other processes. In contrast, Charon, if it finished losing its methane early on, should have a much fuller cratering record preserved on its surface.

Like Triton, Pluto's relatively low abundance of water compared to rock argues for accretion in the cold outer region of the solar nebula. However, whereas Triton was subsequently captured by a major planet, Pluto avoided this fate. Instead, Pluto has wandered until its orbital period has become locked in a stable 3:2 orbital resonance with Neptune's orbit about the Sun. Whether heating events in Pluto, and indeed Charon, were ever able to produce a variety of terrain types similar to that seen on Triton remains unknown, because although intriguing, the surface albedo patterns are seen at too low a resolution to be informative, as comparison with our own Moon at comparable resolution shows (fig. 8.3). The issue will remain unresolved until a spaceprobe is sent to Pluto. As you will see in the final chapter, the earliest arrival date for this is 2012.

8.3 THE KUIPER BELT AND BEYOND

As reported in chapter 2, it has been confirmed that Pluto and Charon are not the only substantial bodies beyond Neptune. By 1999 over a hundred Kuiper belt objects had been discovered, mostly in near-circular orbits with semimajor axes of between 30 and 50 AU. A total undiscovered population of some seventy thousand Kuiper belt objects more than 50 km in radius can be inferred, assuming most orbits lie within 20° of the ecliptic. Only the very largest are likely to be near spherical and so qualify as "worlds" in the terms of this book, but these may number in the hundreds. We have too little information for informed speculation about their geology, but it seems unlikely that they could have experienced sufficient heating to have allowed them to differentiate. Therefore, they are likely to be passive, heavily cratered worlds. Being so small and distant, Kuiper belt objects are extraordinarily difficult to observe and to determine surface compositions of spectroscopically. However, they can be classified into two groups on the basis of general color: a red group and a spectrally neutral group. The reddening may indicate the presence of tholins or other products of long-term radiation damage in hydrocarbon-bearing ices, which is obscured in the other class by temporary volatile frosts or fresh material distributed across the surface by impacts. Methane has been tentatively identified in some spectra. As you will see in chapter 9, there is a chance that a Pluto flyby probe may be able to proceed onward for an encounter with a Kuiper belt object around the year 2020. Until then, our best chance of learning more about these objects may be to study the apparently similar Centaur class of icy asteroids in the Saturn–Uranus region that appear to be Kuiper belt objects that have been scattered inward, and the two outer irregular satellites of Uranus (discovered in 1997) that resemble the reddened group and could be captured Centaurs.

Each of the well-observed Kuiper belt objects closer than about 40 AU is in orbital resonance with Neptune, such that the ratio between its orbital period and Neptune's orbital period is a ratio of small, whole numbers (as we have seen, the ratio is 3:2 in the case of Pluto). Thus, they can avoid colliding with Neptune, but chance, chaotic interactions may occasionally send one inward to become a Centaur or indeed outward into eccentric orbits extending beyond the Kuiper belt proper. The presence of an outer "scattered" Kuiper belt began to be indicated in 1996 by the discoveries of the first two bodies of that class: $1996RQ_{20}$ and $1996TL_{66}$, whose radii are estimated to be about 150 km and 250 km, respectively. Both have eccentric orbits taking them well beyond the classical Kuiper belt, as far as 130 AU at aphelion in the case of $1996TL_{66}$. The total mass of bodies in the "scattered" Kuiper belt probably exceeds that in the classical Kuiper belt, in which case there is a host of worlds out there awaiting the attentions of future generations of geologists.

We shall turn in the final chapter to a summary of the most important things we need to find out to understand the geology of our companion worlds in the outer solar system, and the missions that are planned to go at least some way toward achieving this.

9   What Next?

Come my friends,
'Tis not too late to seek a newer world
Tennyson,
Ulysses

It is in the nature of science always to want to know more. This trait is certainly no less marked in the planetary sciences than in other disciplines. Ten years after the first Moon landing (in 1969), we still had very little information on the moons of the outer planets. Today we have maps of large parts of most of them and can speculate about their composition and evolution using the landforms, topography, and spectra of their surfaces, supported by cosmochemical models. However, there is still a great deal that we need to find out. Many of the present interpretations are probably flawed in detail, and some are certain to be down-right wrong.

9.1 THE MISSING DATA

Clearly, the satellites of the outer planets have a lot to tell about the variety of styles in which planetary bodies with different thicknesses of lithosphere and different histories of internal heating can evolve. What, then, do we most need to determine if we are to understand the origin and evolution of the outer planet satellites, and to take advantage of the light they can cast on planetary bodies in general? As is apparent in the preceding chapters, each of these worlds has its own characteristics, so we are faced with trying to understand the histories of at least nineteen sizable bodies. Generalizations are unwise, and it is dangerous to extrapolate from one body to another.

One thing that would help enormously would be if we could analyze samples collected from the surfaces of some of the outer planet satellites. The evidence for methane and ammonia in many of the satellites of Saturn and Uranus is circumstantial, and although methane has been found spectroscopically on Triton and Pluto, ammonia has yet to be detected in the reflectance spectrum of *any* icy body in the outer solar system. It seems inescapable that ammonia is there, but no one has proved it yet. Furthermore, we have only a poor concept of the relative importances of silicates and carbonaceous material in the dense fractions of these bodies. Sampling of surface material is probably several decades away (except for Titan and possibly Europa), and the most we can hope for is better remote analysis using improved surface spectra obtained by orbiters, flyby probes, and Earth-based observations.

The mystery of the composition of the dark material on Iapetus, and whether the opposite hemisphere's dark-floored craters contain the same material as that blanketing the dark hemisphere, is particularly intriguing. If this could be solved by spectroscopic imaging, it would tell us a considerable amount about both Iapetus's surface and its interior, and the possibility of a volcanic origin for the source of the dark material.

To get a good idea of the internal structure of a planetary body, it is necessary to deploy seismometers on its surface. By examining the nature and relative timing of vibrations that have propagated from sources of seismic disturbance, such as "earthquakes" and meteorite impacts, internal phase changes and other interfaces between layers of different density can be located, densities can be inferred, and the presence of liquid or partially molten zones can be deduced from the attenuation or nontransmission of shearing waves (as opposed to pressure waves). The possibility of these sorts of experiments in the outer solar system seems a long way off at present (so far, apart from on Earth, they have been achieved only on the Moon and, very crudely, on Mars). A less ambitious option is to use detailed measurements of a satellite's gravity field, derived from perturbations of the trajectory of a probe in a close flyby, to infer internal density variations. The value of this approach, backed up by magnetic measurements, is demonstrated by the refinements in our models for the interiors of the galilean satellites as a result of the *Galileo* mission. However, Callisto's magnetic field is perplexing; it would seem to indicate a subsurface ocean and yet the surface appears to be very ancient. Maybe further analysis of *Galileo* magnetometer data will clarify how close to the surface this putative ocean needs to be to explain the magnetic field.

As for the imaging of satellite surfaces, except for Europa, Ganymede, and Callisto we badly need fuller, near global coverage at a resolution of about 1 km or finer to prevent us from making inappropriate generalizations about an entire body based on good pictures of only a fraction of its surface. An object lesson in this respect is the way our perception of Mars as a Moon-like cratered planet, based on the low resolution and patchy coverage obtained by *Mariner-4* in 1965, was changed by the better and more extensive mapping by *Mariner-9* (1971) and the two *Viking* missions (1976) to that of a varied world with large volcanoes, canyons, and tracts of sand dunes. We also need at least local coverage by more detailed images to enable smaller details to be picked out, like the glorious high-resolution views of parts of Europa, Ganymede, and Callisto returned by *Galileo*. It would also be useful to obtain pictures of coincident areas seen from different directions. This would provide stereoscopic images that would enable the surface topography to be mapped, overcoming many of the limitations of shadow-length measurements and shape-from-shading techniques, which are all that is usually available at present.

By way of example, more extensive and improved imaging of Enceladus would enable us to complete the global tectonic map and to assess the extent, if any, of recent or present-day activity. Rhea is another world where the *Voyager* coverage is frustrating. In this case the anti-Saturn hemisphere was poorly imaged (50 km resolution), just good enough to show pale wispy markings that might be the former sites of major tectonic activity, on a world where models of its thermal evolution suggest that we should expect some signs of geological events to have survived. Titan is in a class of its own: only radar images would be capable of revealing details of its tectonic and volcanic history, and its surface morphology may contain evidence of changes in its climate linked to the "greenhouse" effect and other factors.

We are also lacking complete global views of the satellites of Uranus and Neptune. Among these, Titania's evidently complex tectonic history makes the low resolution of the images of it particularly frustrating. There is no immediate prospect of remedying this situation, but a mission is on its way to Saturn that should greatly improve our understanding of its attendant satellites, and there is also provisional funding for probes to Europa and Pluto. These are now briefly described.

9.2 CASSINI

The *Cassini* mission to the Saturn system is a joint enterprise between NASA and the European Space Agency. Named after the discoverer of the widest gap in Saturn's rings, *Cassini* is probably the last of the old-style big space missions, having twice the mass of *Galileo* at 5548 kg. It was launched on 15 October 1997 and is due to arrive at Saturn in 2004, after two gravity-assist flybys of Venus and one of Earth in 1998–99 (gaining 6–7 km s^{-1} of speed each time) and one of Jupiter in 2000, picking up a further 2 km s^{-1}. This is described as a Venus-Venus-Earth-Jupiter gravity assist (VVEJGA) trajectory. Unfortunately, there are no plans to recover any images while the probe passes through the Jupiter system.

Cassini will pass 20,000 km above Saturn's cloud tops in July 2004, when it will fire one of its engines to slow itself sufficiently to be captured into an elongated orbit. *Cassini* will round the planet and coast away to a distance of 178 Saturn radii before a further engine burn to raise the near point (periapsis) of its orbit to 8.2 Saturn radii, thereby enabling repeated close encounters with Titan.

On 8 November 2004 *Cassini* will detach a 318 kg probe named *Huygens* and send it ahead to enter Titan's atmosphere on 27 November. *Huygens* will be slowed down first by a heat shield and then by parachute, and should hit the surface at a speed of 5–6 m s^{-1} about two hours after entering the atmosphere. During its descent *Huygens* will analyze atmospheric gases and aerosols and record downward- and sideways-pointing images in the visible and infrared. If all goes well, there should be sufficient battery power remaining for at least half an hour of operation on the surface. The first bit of the probe to touch Titan's surface will be a penetrometer intended to test the strength of any solid surface it hits. If the surface is liquid, a densitometer will come into play to determine the methane-to-ethane ratio, and an acoustic sounder will be used to determine its depth. With luck, the on-board accelerometer will be able to measure any rocking motion caused by waves.

Although it may get a chance to image Phoebe from a range of 52,000 km during its initial approach to Saturn, the *Cassini* orbiter will really come to life only after relaying the *Huygens* data back to Earth. Its first flyby of Titan will be on 28 November, and over thirty more are scheduled during the sixty orbit, four year primary mission, mostly at a range of 950–2500 km and giving an excellent opportunity to map Titan's internal mass distribution and magnetic field. The orbital pattern will be petallike, similar to *Galileo*'s at Jupiter (see fig. 3.9), except that Titan gravity assists will be used to vary the orbital inclination between 0° and 75°. This will allow thorough mapping of Saturn's magnetosphere, and polar as well as equatorial imaging of Saturn and Titan. About 20–30 percent of Titan's surface will be imaged using a synthetic aperture radar to penetrate the atmosphere, with a best resolution of about 500 m. The potential fruits of such an instrument package were demonstrated by the *Magellan* probe, which produced spectacular radar images of the surface of Venus in 1990–91.

Other, more conventional imaging and visible-infrared spectrometers will cover Saturn itself and its other satellites. There will be flybys of Rhea, Dione, Iapetus, and Enceladus at less than 1000 km and twelve further encounters within 50,000 km of these four plus Tethys and Mimas, which should in total, offer a considerable advance on what we know from *Voyager*. In particular, *Cassini* offers our best foreseeable chance to find out whether or not Enceladus is currently active and to discover the composition and origin of the dark material coating the leading hemisphere of Iapetus.

The *Cassini* mission is due to terminate at the end of June 2008, four years after arriving at Saturn. However there is every reason to hope that, like the *Voyagers* and *Galileo*, the *Cassini* orbiter will remain in good health. If so there are several options for an extended mission, probably operating on a reduced level of funding like the *Galileo Europa Mission*. By continuing in a similar mode, *Cassini* could extend its radar mapping of Titan and its conventional imaging of the other satellites. Other options are to direct it closer to Saturn and its ring system, or to "aerobrake" it into orbit around Titan which would give information on atmospheric density as well as allowing more prolonged, complete and detailed radar mapping.

9.3 *EUROPA ORBITER*

Europa is an obvious priority target for further exploration, because of its recently active surface, the likelihood of a water ocean at relatively shallow depth beneath the ice, and even the possibility of life subsisting on hydrothermal energy. There have been proposals to look for signs of life by timing the impact of a projectile onto Europa's surface so that a shallow flyby or orbiter probe could fly through the ejecta curtain and collect a few nanograms for analysis (unusual ratios between stable isotopes of oxygen or carbon would be strong signs of biological activity). Others have suggested landers capable of drilling or melting a path through the ice to release a tethered or free-swimming robotic submersible (a "hydrobot") that could explore the ocean. These ambitious goals may be achieved in our lifetimes, but first it is likely that a more modest but still technologically sophisiticated low-mass probe named *Europa Orbiter* will explore Europa as part of NASA's Outer Planets/Solar Probe Program.

The *Europa Orbiter* will probably be the first mission to orbit an outer planet satellite. Its primary science goals will be to identify sites of recent or current activity on the surface, to determine whether or not there still is a subsurface ocean, and if so, to characterize the three-dimensional distribution of subsurface liquid water and the overlying ice layers. In order to meet these objectives, in addition to an imaging system it will carry an ice-penetrating radar intended to transmit microwaves through the ice and detect their echo from the ice/water interface, and probably a laser altimeter to compare Europa's actual shape with its tidally predicted shape.

Launch is tentatively scheduled for late 2003 on a direct trajectory to Jupiter, with arrival between three and four years later. Insertion into Jupiter orbit will be followed by a *Galileo*-like orbital tour lasting up to two years. This will allow opportunities to image other satellites, but its main purpose will be to adjust the spacecraft's trajectory in preparation for a five month series of close Europa encounters and engine burns necessary to achieve polar orbit around Europa at an altitude of between 100 and 300 km. During a nominal orbital life of one month, the *Europa Orbiter* should produce global color images at a resolution of better than 300 m per pixel, and monochromatic images of all feature types at about 10 m per pixel. Its ice-penetrating radar is intended to produce a global map of ice thickness with a depth resolution of about 100 m. One possible shortcoming is that this is unlikely to be able to detect the ocean if it lies at more than about ten kilometers depth, and the detection limit may be shallower if the ice is rich in impurities such as salts.

9.4 PLUTO-KUIPER EXPRESS

The other outer solar system mission planned under NASA's Outer Planets/Solar Probe Program is the long-awaited mission to Pluto (and beyond). Its first objective will be to complete the reconnaissance phase of our planetary exploration by achieving the first close-up observations of the Pluto–Charon system, with the option of continuing to an encounter with one or more Kuiper belt objects if the craft remains healthy.

Pluto-Kuiper Express will use lightweight advanced technology hardware and innovative computer software in order to meet the stringent mass and power constraints of such a mission. It is scheduled for launch at the end of 2004 but could go a year earlier if it is ready before the *Europa Orbiter*. The route to Pluto will probably be a Jupiter gravity assist (JGA) trajectory, with Pluto–Charon flyby after eight or more years flight time depending on mass, launch vehicle, and exact trajectory. The flyby speed at Pluto will be about 18 km s^{-1}, with a closest approach of 15,000 km. It is hoped to obtain panchromatic images of the sunlit regions of Pluto and Charon at a resolution of about 1 km per pixel, to record infrared spectra for mapping surface composition at 5–10 km per pixel, and to determine the composition and structure of Pluto's atmosphere. While *Pluto-Kuiper*

Express is hurtling onward into the Kuiper belt, it will take several months to send back to the Deep Space Network the gigabit or more of science data recorded during the encounter.

9.5 BOLDLY GOING . . .

Thus, despite the perennial threat of financial cutbacks, we are likely to continue to amass a considerable volume of new data on several icy worlds over the coming decades. However, there seems to be little prospect of a prompt return to Uranus and Neptune, so the mysteries of Triton's climatic changes and the volcanism and tectonism on Miranda, Ariel, and Titania will remain unresolved for now. Costs, potential scientific returns, and the increased risk of malfunction during a long-duration mission must all be weighed when assigning priorities for exploration. This means that the delights of the outer solar system have to be balanced against what could be achieved by more detailed exploration of objects nearer to home, such as the Moon, Mars, and the asteroids, where there may even be economic paybacks in the medium to long term. It is really only the (possibly illusory) lure of finding life that has enabled NASA to allocate funds for a return to Europa.

There are exciting times ahead, but we have probably passed the end of what others before me have called 'the golden age of solar system exploration', during which we gained our first views of so many worlds never before seen in any detail. The *Galileo* mission was a triumph despite its technical hitches, and if *Cassini* performs as intended it will teach us a lot, too. However, history may well regard as the climax of this era the marvelous years from *Voyager-1*'s Jupiter encounter in 1979 until *Voyager-2*'s Neptune–Triton flyby in 1989, when the surfaces of no fewer than sixteen satellites greater than about 200 km in radius were charted for the first time. For planetary geologists there is unlikely to be such an eye-opening episode again until we see our first detailed images of the planets of another star. But that will be another story.

Appendix

Names of Surface Features

The names of features on the outer planet satellites are assigned by the International Astronomical Union according to the following conventions:

SATELLITES OF JUPITER

Io   Gods and heroes associated with fire, sun, thunder, or volcanoes; also people and places associated with the Io myth or from Dante's *Inferno*.

Europa   Celtic gods, heroes, and myths; people and places associated with the Europa myth; places in ancient Egypt.

Ganymede   Gods, heroes, and places from ancient Egypt and the fertile crescent.

Callisto   People and places from northern myths.

SATELLITES OF SATURN

Mimas   People and places from Sir Thomas Malory's *Le Morte d'Arthur.*

Enceladus   People and places from Sir Richard Burton's version of the *Arabian Nights.*

Tethys   People and places from Homer's *Odyssey.*

Dione   People and places from Virgil's *Aeneid.*

Rhea   People and places from African, Asiatic, and South American creation myths.

Titan   Not yet mapped, though names are likely to come from myths involving the Titans.

Iapetus   People and places from the medieval French epic *Chanson de Roland* (Song of Roland).

SATELLITES OF URANUS

Miranda	Humans from Shakespeare's *The Tempest*; Shakespearean place names.
Ariel	Bright (good) spirits from worldwide mythologies.
Umbriel	Dark (evil) spirits from worldwide mythologies.
Titania	Minor Shakespearean female characters.
Oberon	Shakespearean heroes.

SATELLITES OF NEPTUNE

Triton	Aquatic gods and places.

PLUTO AND CHARON

Not yet mapped, though names are likely to be connected with underworld deities.

Glossary

accretion The formation and growth of planetary bodies by collision with other bodies and gravitational sweeping up of gas and dust.

accretional heating Heating of a planetary body as it grows by the kinetic energy liberated when smaller objects collide with it. Most of this heat is generated near the temporary surface, but it can be carried inward by conduction.

aerosol Liquid droplets small enough to be supported in an atmosphere. Titan's atmosphere is opaque to visible light because of aerosols of ethane and other, more complex organic molecules.

albedo A measure of the brightness or reflectivity of a surface. An albedo of 1 means that all the incident light is reflected by a surface, so that it appears white. (In this book, the values of albedo quoted are of "normal" or "geometric" albedo calculated for light falling perpendicularly on the surface, for small areas, or along the line of sight, for spheres.) For an icy moon, an albedo of nearly 1 means that the ice is clean. A low albedo (ranging down to less than 0.1) implies contamination of the ice by carbonaceous or **silicate** dust particles, **tholins**, and/or rock fragments; this could be because the outer part of the body accreted homogeneously and did not subsequently melt or differentiate in any way, or because the surface has picked up a coating of meteoritic dust over the ages. If the ice contains methane, it could also darken with age by exposure to cosmic radiation and solar ultraviolet light. The grain size of the ice also has an effect, smaller grain sizes leading to a higher albedo.

asthenosphere A weak layer below the **lithosphere** of a terrestrial planet or icy moon, which behaves as a fluid on a geological timescale, and which is capable of solid-state **convection** in response to the internal generation of heat at a rate sufficient to transport heat outward at least as efficiently as conduction.

AU (astronomical unit) A convenient unit to measure interplanetary distances, defined as the mean distance of Earth from the Sun, that is, 150 million kilometers (93 million miles).

basalt A type of volcanic rock, with comparatively little silica in its makeup, and of low viscosity. Earth's oceanic crust is formed of rock of basaltic composition.

billion A thousand million (10^9).

brine A fluid produced by **partial melting** of a mixture of **ice** and **salt**.

caldera A crater formed largely by collapse of its floor, when an underlying magma body is removed. This can happen by drainage of magma back to depth or by eruption.

carbonaceous chondrites Meteorites formed of **chondrules** in a matrix of hydrated minerals that also contains organic (carbon-bearing) molecules. Their bulk composition is thought to represent that of the nonvolatile component of the **solar nebula**.

chondritic meteorites Primitive meteorites consisting of agglomerations of **chondrules**.

chondrule A silicate globule, about 1 mm to 1 cm in diameter, commonly found in primitive meteorites, which is believed to represent a droplet formed by condensation from the **solar nebula**.

clathrate Ice of one compound holding molecules of another at otherwise void spaces in its crystalline lattice. Ice I is capable of holding one molecule of methane (CH_4), carbon monoxide (CO), or nitrogen (N_2) for every six molecules of water (H_2O).

convection The transfer of heat in a fluid by means of circulation, which is driven by a strong temperature gradient.

core The inner part of a diffentiated planetary body, as distinct from its **mantle**. In an icy body the core is essentially rocky but may contain a metallic (iron-rich) inner core.

crust Any outer, chemically and mineralogically distinct part of the lithosphere of a planetary body, which overlies the **mantle**.

cryovolcanism Volcanism on an icy body, where the molten material is produced from **ice** rather than **silicates** as on Earth.

diapir A geological structure formed when more-dense material (on Earth this is often rock) is underlain by a layer of deformable, less dense material (on Earth this is often "rock salt," sodium chloride). Buoyancy forces cause the less dense material to collect into pods up to several tens of kilometers in size that rise through the overlying layer and may pierce through to the surface.

differentiation A gravity-driven process whereby a planet or satellite becomes layered. Denser phases segregate inward and lighter phases work their way outward.

eccentricity A measure of the departure of an orbit from circularity; the greater the eccentricity, the more strongly pronounced the elliptical nature of the orbit.

ecliptic The plane of Earth's orbit about the Sun. The orbits of most of the other planets are close to the same plane, the exception being Pluto, whose orbit is inclined over 17°.

fumarole A vent from which volcanic gases escape, usually indicated by condensation around the vent (on Earth, often in the form of sulfur) and/or chemical alteration and discoloration of the surface material.

graben A valley formed by the down-dropping of its floor along steep faults that define the valley walls.

heterogeneous accretion Formation of planetary bodies by sequential condensation and **accretion**, beginning with the most **refractory** elements and continuing through progressively more **volatile** elements. If this happened, then the planetary bodies actually formed with a layered structure, although this is likely to have been modified by subsequent events.

homogeneous accretion A model for the formation of planetary bodies by the **accretion** of **refractory** and **volatile** elements at the same time. If this happened, then any layering within a planetary body has to be due to subsequent processes of **differentiation**.

horst An uplifted strip of terrain, bounded by faults on either side, or a block left standing high as a result of the down-dropping of **grabens** on either side.

ice In the outer solar system, any frozen **volatile** or mixture of volatiles, most abundantly water, but also including ammonia, methane, carbon monoxide, carbon dioxide, and nitrogen.

igneous An adjective describing a rock that has previously been molten. If it was erupted onto the surface, then it is a volcanic igneous rock; otherwise, it is intrusive.

ignimbrite A **pyroclastic** rock consisting of silica-rich fragments, often welded together and believed to be formed by the flow of rock fragments away from the site of an explosive eruption. The entrainment and heating of the surrounding air make the flow buoyant and able to travel vast distances at high speed.

infrared Radiation the wavelength of which is greater than that of visible light. Near-infrared extends from 700 nm to about 3 μm in wavelength. The thermal infrared is from about 3 μm to about 30 μm and is the region where matter at normal temperatures emits radiation most strongly as a result of its heat.

isostasy The operation of buoyancy forces on the outer, rigid layer of a planetary body. If this layer becomes thickened at the surface, the base of the layer will tend to subside if there is an underlying weak layer to accommodate the movement. The process stops when the weights per unit area of all columns of rock extending to the interior of the planet have become equal. Isostasy can operate only if the strength and thickness of the outer rigid layer are low enough.

Kuiper belt A region in the outer solar system (close to Neptune's orbit and beyond) occupied by icy bodies (Kuiper belt objects) ranging in size from a few hundred kilometers downward.

lag deposit A residue of nonvolatile material on a surface, left when a more volatile component (with which the nonvolatile component was originally mixed) has been removed.

late heavy bombardment The earliest age of cratering on the Moon of which good evidence remains (ending about 3.9 billion years ago), older traces having been obliterated by subsequent impacts. The late heavy bombardment appears to represent the tailing off of the bombardment of the planets by debris left over from the formation of the solar system, as this debris became virtually exhausted. This event is well documented in the inner solar system by the radiometric dating of lunar samples, but it cannot be proven that the latest phase of heavy bombardment preserved within the satellite systems of each outer planet occurred at the same time, or that it is the result of the same population of impactors.

limb The sunlit edge of the visible disk of a **planetary body**.

liquidus On a phase diagram of pressure against temperature, a line showing the conditions under which melting occurs.

lithosphere The rigid outer layer of a terrestrial planet or icy moon, which overlies the **asthenosphere** (if present).

magma Molten rock, especially when occurring in a body rising toward the surface (when erupted as a liquid it is known as lava). The term can be generalized to include melts within an icy moon.

magnetosphere The region of space surrounding a planet in which the planet's magnetic field dominates over that of the solar wind.

mantle The chemically distinct layer lying between the core and the **crust** of a planetary body. The mantle may be said to extend to the surface if there is no chemically distinct crust. Note that mantle and crust are chemical terms, whereas **lithosphere** and **asthenosphere** are defined by differences in physical properties.

moon Used here to mean any natural satellite of a planet. The name of Earth's satellite, the Moon, begins with a capital letter.

newtonian Regarding a fluid, meaning that the rate at which it deforms is proportional to the applied shear stress divided by its viscosity. Non-newtonian fluids have a **yield strength** that must be overcome before they will deform.

nucleosynthesis The creation of heavier elements by the fusion of lighter elements during nuclear fusion reactions within a star.

occultation When one astronomical object hides another by coming directly between it and the line of sight of the observer. See also **radio occultation**.

orbital resonance The situation when two or more bodies orbiting the same larger body have orbital periods that are a simple ratio of one another, for example, Io, Europa, and Ganymede's 1:2:4 orbital resonance about Jupiter, and Neptune and Pluto's 2:3 orbital resonance about the Sun. Such situations are brought about by mutual gravitational and tidal interactions and, once established, tend to be stable over long time periods.

palimpsest Usually a bright circular feature with little or no detectable relief, most common on Ganymede but also known on Callisto. Typically a hundred or more kilometers in diameter, they are thought to represent the trace of a large impact that occurred when the **lithosphere** was still thin, the crater form having been destroyed either by viscous **relaxation** or the extrusion of warm ice or slush.

partial melting The phenomenon whereby a mixture of **silicate** minerals, a mixture of **ices**, or a mixture of ice and **salts** melts at a lower temperature than would any one of its components in isolation. The initial melt has a chemical composition differing from that of the solid, and (especially if it is less dense than the solid) may rise toward the surface to produce a volcanic eruption.

pixel A picture element in a digital image, the building block out of which an image is composed.

planetary body Used here to denote a planet, one of its satellites, or any other sizable body.

planetesimal An object from millimeters up to maybe a hundred kilometers in diameter formed during an intermediate stage of the formation of the solar system, by cohesion between the tiny grains that were formed by condensation from the **solar nebula**. Subsequently, the planets and satellites formed by gravitational **accretion** of planetesimals.

plate tectonics The generally accepted description of global **tectonics** on Earth, which says that the **lithosphere** is divided into several rigid plates. These are spreading apart at rates of up to about 10 cm per year at oceanic spreading axes (midocean ridges), and converging at continental collision zones and subduction zones beneath island arcs and Andes-type mountain ranges where lithosphere is destroyed at a rate sufficient to balance its creation at spreading axes.

Population I Originally defined at Saturn, and applied by some to Uranus, this refers to the oldest suite of craters preserved in the satellite system (and the impactors that produced these). It is generally thought to represent the sweeping up of postaccretional debris, although not necessarily the same as the **late heavy bombardment**.

Population II As for **Population I**, but a later phase of cratering, producing a smaller proportion of large craters. It is possibly the result of collisions with debris in orbit around the same planet. Such debris could have resulted from the destruction of a satellite because of a major impact. Satellites that preserve reasonable evidence of geological activity generally have terrains, or even their entire surface, that were **resurfaced** after the end of the Population I bombardment and are dominated by Population II craters alone. The Population II flurry of impacts is unlikely to date from the same time at each planet.

Population III The gradually declining background flux of impacts (by cometary material and other debris) continuing to the present day, which is sometimes distinguished from **Population II** (which is thought to represent one or more shorter flurries).

prograde Referring to the orbital motion of a planet when it is anticlockwise as seen from above the plane of the solar system (above the Sun's north pole) and, similarly, of a satellite when its orbit is anticlockwise as seen from above its planet's north pole. It is also used for rotation of a planet or satellite on its axis that is anticlockwise as seen from above the body's north pole. Most bodies in the solar system have both prograde orbits and prograde rotation. The opposite is **retrograde**.

protoplanet In models for solar system formation and development, a large **planetesimal** that, because of its larger gravity field, grows faster than the others in its vicinity. Eventually it reaches planetary size, **differentiates**, and becomes a planet.

protosatellite disk A disk of gas and dust around a planet, or **protoplanet**, from which its satellites formed.

pyroclastic A term describing a volcanic eruption and its resulting deposits when it is caused by explosive fragmentation of the molten rock (or ice) as a result of the pressure built up by gases escaping from or through the melt.

radiogenic heating Heating of a planetary body from within by heat generated in the decay of radioactive elements. The short-lived isotope ^{26}Al was probably very important during the first few million years of the solar system's life, but subsequently most radiogenic heat has originated in much longer lived isotopes of uranium, thorium, and potassium. These elements reside within silicates, and there is no significant source of radiogenic heat within ices except perhaps in the form of potassium **salts**.

radiometric dating A method of working out the time since a rock or mineral grain formed by measuring the isotopes of elements involved in radioactive decay.

radio occultation A technique used to determine a profile of the density of the atmosphere with respect to altitude for a planetary body, by measuring the way in which a radio signal from a space probe is weakened as the signal passes through the atmosphere while the space probe passes behind the planetary body as seen from Earth. These density data give information on the ratio between temperature and the mean molecular weight of the atmosphere, so if temperature can be determined independently, say by **infrared** techniques, radio occultation data can be used to determine the mean molecular weight of the atmosphere.

refractory A relative term, denoting the earliest substances to condense from a cooling vapor (as in the **solar nebula**), while the temperature was still high.

regolith A layer of fragmentary debris on the surface of a solid body, produced by meteoritic impacts.

relaxation The process whereby a **lithosphere** deforms under its own weight, in response to major topographic features such as large craters or **grabens**. The result is not a flat surface, but one on which the relief is subdued. It occurs in lithospheres that are thin and/or warm.

reseau marks An array of calibration dots on the faceplate of a vidicon camera, which are put there to enable electronic geometric distortions to be corrected. If reseau marks are not synthetically removed from an image they appear as black dots.

resolution A measure of the size of the smallest detail visible on an image. In this book the values of resolution quoted represent the size of the pixels; elsewhere resolution is sometimes quoted in terms of kilometers per line pair.

resurfacing Creation of a new surface on a planetary body by **volcanic** or **tectonic** processes. These can include burial of older features under the products of eruptions and the breakup of old **terrains** by faulting.

retrograde Referring to the orbital motion of a satellite if it is clockwise as seen from above its planet's north pole; most bodies in the solar system have orbits that are the opposite, or **prograde**. It is also used for rotation of a planet on its axis that is clockwise when seen from north of the plane of the **ecliptic**.

rheological Pertaining to the way a material responds to stress applied at different rates, in particular, how it can flow in the solid state under high pressure.

salt Any of a wide range of simple compounds, notably sulfates, carbonates, and chlorides of metals such as magnesium, sodium, and potassium, that are expected contaminants in solar system **ice** as a consequence of chemical reactions involving water and rock.

shear strain The amount by which a fluid (or any other substance) is deformed in response to a **shear stress**.

shear stress The force tending to deform something by shearing one layer over another.

silicate A silicate mineral is any of the common rock-forming minerals on Earth and the other terrestrial planets, consisting of silicon and oxygen, usually with a mixture of iron, magnesium, sodium, potassium, calcium, and/or aluminium. The term "silicates" is often used to mean minerals or rocks in general.

solar nebula The cloud of gas and dust around the proto-Sun from which the solar system formed.

sputtering The vaporization and molecular breakdown of ice as a result of impact by micrometeorites or energetic cosmic ray particles. This proceeds at extremely slow rates on the icy satellites, but it could help to increase the concentration of rock fragments on the surface. See also **sublimation**.

strain rate The rate at which a material is made to deform.

strike-slip A geological term referring to faults in which the dominant movement is lateral, so as to displace features sideways from their original positions.

sublimation The process whereby ice, in a vacuum, evaporates directly from the solid to the vapor state. This occurs extremely slowly at the low temperatures prevalent in the outer solar system, but over the age of the solar system it could provide, by removal of the ice, a possible mechanism for concentrating a residual **lag deposit** of rocky material on the surface of an icy satellite. See also **sputtering**.

synchronous rotation Rotation of a planetary satellite in the same period as its orbit, so it always keeps the same face toward the planet. This is brought about through tidal interactions and is also known as "captured rotation."

tectonics A geological term referring to processes of faulting or other distortion or disruption of a body's **lithosphere**, often on a global scale, and almost always as a result of large-scale internal movements below the lithosphere.

terminator The boundary between the sunlit and dark parts of the surface of a **planetary body**.

terrain A tract of surface that has a set of recognizable characteristics, indicating that it probably has a different history from that of other terrains on the same body.

terrestrial Referring to something that occurs on Earth; also, in the term "terrestrial planet," meaning any of the planets whose outer layers are dominated by silicates, that is, Mercury, Venus, Earth, the Moon, and Mars (and, according to some definitions, Io).

tholins Large class of solid, tarlike nonvolatile substances produced in ice or hydrocarbon-rich atmospheres when methane and other constituents react under the influence of charged particles or ultraviolet radiation.

tidal heating (tidal dissipation) Heating of a satellite as a result of varying deformation due to the action of its planet's gravity on a tidal bulge.

transcurrent fault A simple **strike-slip** fault in which the relative motion between units on either side is the same as the visible sense of displacement between features offset by the fault (compare **transform fault**).

transform fault In Earth's oceans a sideways offset between two segments of a spreading axis. As a result of the axial spreading, a transform fault across which the spreading axis is offset to the right is actually a site of relative motion to the left.

viscosity A measure of the ease with which a fluid flows; it can be defined as **shear stress** divided by rate of **shear strain**. It is conventionally measured in poise (1 poise = 0.1 kg m^{-1} s^{-1}). Typical values of viscosity are water, 10^{-2} poise; honey and $NH_3 \cdot 2H_2O$ (ammonia hydrate melt), 100 poise; sulfur, about 0.1 to nearly 10^3 poise according to temperature; terrestrial basalt lava, 10^3 poise; terrestrial rhyolite lava, 10^6 poise; and pure ice at 240 K, 10^{16} poise.

viscous relaxation see **relaxation**.

volatile In cosmochemistry, a relative term denoting substances that do not condense from a cooling vapor (as in the **solar nebula**) until the temperature has dropped very low. Used in volcanology to denote an abundant chemical species in the vapor phase that escapes from **magmas** and may cause an explosive (**pyroclastic**) eruption.

volcanism The eruption of molten material or gas-driven solid fragments at the surface of a **planetary body**. On icy moons, the phenomenon is sometimes called **cryovolcanism** to distinguish it from silicate or sulfur volcanism.

yield strength The property of a non-newtonian fluid, such that a minimum **shear stress** must be applied to it before it will begin to flow. Terrestrial lavas and icy **magmas** all have yield strengths.

Bibliography

I wrote this book because I could not find one about the geology of the outer solar system at a similar level. If you want a broad overview of planetary satellites as a whole, with considerable technical detail, you should consult the space science series from the University of Arizona Press, Tucson:

Bergstralh, J. T., Miner, E. D., and Matthews, M. S. (ed.) (1991). *Uranus*, 1076pp.

Burns, J. A., and Matthews, M. S. (ed.) (1986). *Satellites*, 1021pp.

Cruickshank, D. P. (ed.) (1995). *Neptune and Triton*, 1249pp.

Stern, A. S., and Tholin, D. J. (eds.) (1997). *Pluto and Charon*, 728pp. and the dated but still-useful:

Morrison, D. (ed.) (1982). *Satellites of Jupiter*, 972pp. University of Arizona Press, Tucson.

Gehrels, T., and Matthews, M. S. (ed.) (1984). *Saturn*, 968pp. University of Arizona Press, Tucson.

A book at a generally less detailed level than this one, but dealing with geology on the terrestrial planets as well as planetary satellites, is:

Greeley, R. (1994). *Planetary landscapes*, 2nd ed., 275pp. Chapman and Hall, London.

A readable review of the whole solar system is:

Beatty, J. K., Petersen, C. C., and Chaikin, A. (1999). *The new solar system*, 4th ed., 421 pp. Sky, Cambridge, Massachusetts, and Cambridge University Press, Cambridge.

Techniques used in mapping planetary landscapes and the conventions used for naming features are discussed at a simple level in:

Greeley, R., and Batson, R. M. (ed.) (1990). *Planetary mapping*, Cambridge Planetary Science Series 6, 296pp. Cambridge University Press, Cambridge.

The nature and distribution of ices in the solar system are reviewed at a technical level in:

Schmitt, B., de Bergh, C., and Festou, M. (ed.) (1998). *Solar system ices*, Astrophysics and Space Sciences Library, Vol. 227, 826pp. Kluwer Academic Publishers, Dordrecht.

In writing this book, I have made use of material from several of the above. For those who want to explore the primary scientific literature on large icy satellites, I give below a list sufficient to get you started. This includes the initial reports of each *Voyager* and *Galileo* encounter published in either *Nature* or *Science* (there are usually related articles in the same journal issue that I have not listed), a few seminal papers, and the more useful reports of other sorts. It is not an exhaustive bibliography of all I have drawn from. I do not list conference proceedings and the like, although I found the abstract volumes for the Lunar and Planetary Science Conference held annually at the Lunar and Planetary Institute, Houston, very useful. I list a few key websites at the end.

Allison, M. L., and Clifford, S. M. (1987). Ice-covered water volcanism on Ganymede. *Journal of Geophysical Research*, **92**, 7865–76.

Anderson, J. D., Lau, E. L., Sjogren, W. L., Schubert, G., and Moore, W. B. (1996). Gravitational constraints on the internal structure of Ganymede. *Nature (London)*, **384**, 541–3.

Anderson, J. D., Lau, E. L., Sjogren, W. L., Schubert, G., and Moore, W. B. (1997). Europa's differentiated internal structure: inferences from two Galileo encounters. *Science*, **276**, 1236–9.

Anderson, J. D., Schubert, G., Jacobson, R. A., Lau, E. L., Moore, W. B., and Sjogren, W. L. (1998). Distribution of rock, metals and ices in Callisto. *Science*, **280**, 1573–6.

Belton, M. J. S., et al. (1996). Galileo's first images of Jupiter and the galilean satellites. *Science*, **274**, 377–85.

Cameron, A. G. W. (1988). Origin of the solar system. *Annual Review of Astronomy and Astrophysics*, **26**, 441–72.

Carlson, R. W. et al. (1999). Hydrogen peroxide on the surface of Europa. *Science*, **283**, 2062–4.

Carr, M. H., et al. (1998). Evidence for a subsurface ocean on Europa. *Nature (London)*, **391**, 363–5.

Chyba, C. F., Ostro, S. J., and Edward, B. C. (1998). Radar detectability of a subsurface ocean on Europa. *Icarus*, **134**, 292–302.

Consolmagno, G. J. (1985). Resurfacing Saturn's satellites: models of partial differentiation and expansion. *Icarus*, **64**, 401–13.

Consolmagno, G. J., and Lewis, J. S. (1978). The evolution of icy satellite interiors and surfaces. *Icarus*, **34**, 280–93.

Crater Analysis Techniques Working Group. (1979). Standard techniques for presentation and analysis of crater size–frequency data. *Icarus*, **37**, 467–74.

Croft, S. K., Lunine, J. I., and Kargel, J. (1988). Equation of state of ammonia-water liquid: derivation and planetological applications. *Icarus*, **73**, 279–93.

Cruikshank, D. P., and Brown, R. H. (1993). Remote sensing of ices and ice-mineral mixtures in the outer solar system. In *Remote Geochemical Analysis: Elemental and Mineralogical Composition* (ed. C. M. Pieters and A. J. Englert), pp. 455–68. Cambridge University Press, Cambridge.

Ellsworth, K., and Schubert, G. (1983). Saturn's icy satellites: thermal and structural models. *Icarus*, **54**, 490–510.

Golombek, M. P., and Banerdt, W. B. (1990). Constraints on the subsurface structure of Europa. *Icarus*, **83**, 441–52.

Hall, D. T., Strobel, D. F., Feldman, P. D., McGrath, M. A., and Weaver, H. A. (1995). Detection of an oxygen atmosphere on Jupiter's moon Europa. *Nature (London)*, **373**, 677–9.

Hamilton, D. P., and Burns, J. A. (1994). Origin of Saturn's E ring: self-sustained, naturally. *Science*, **264**, 550–3.

Hapke, B. (1989). The surface of Io: a new model. *Icarus*, **79**, 56–74.

Herrick, D. L., and Stevenson, D. J. (1990). Extensional and compressional instabilities in icy satellite lithospheres. *Icarus*, **85**, 191–204.

Hogenboom, D. L., Kargel, J. S., Consolmagno, G. J., Holden, T. C., Lee, L., and Buyyounouski, M. (1997). The ammonia–water system and the chemical differentiation of icy satellites. *Icarus*, **128**, 171–80.

Ingersoll, A. P. (1990). Dynamics of Triton's atmosphere. *Nature (London)*, **344**, 315–7.

Jakosky, B. M., and Shock, E. L. (1998). The biological potential of Mars, the early Earth, and Europa. *Journal of Geophysical Research*, **108**, 19359–64.

Janes, D. M., and Melosh, H. J. (1988). Sinker tectonics: an approach to the surface of Miranda. *Journal of Geophysical Research*, **93**, 3127–43.

Jankowski, D. G., and Squyres, S. W. (1988). Solid-state ice volcanism on the satellites of Uranus. *Science*, **241**, 1322–5.

Johnson, M. W., and Nicol, M. (1987). The ammonia–water phase diagram and its implications for icy satellites. *Journal of Geophysical Research*, **92**, 6339–49.

Johnson, T. V., Brown, R. H., and Pollack, J. B. (1987). Uranus satellites: densities and composition. *Journal of Geophysical Research*, **92**, 14884 –94.

Johnson, T. V., Veeder, G. J., Matson, D. L., Brown, R. H., Nelson, R. M., and Morrison, D. (1988). Io: evidence for silicate volcanism in 1986. *Science*, **242**, 1280–3.

Kargel, J. S. (1991). Brine volcanism and the interior structures of asteroids and icy satellites. *Icarus*, **94**, 368 –90.

Kargel, J. S. (1992). Ammonia-water volcanism on icy satellites: phase relations at 1 atmosphere. *Icarus*, **100**, 556–74.

Kargel, J. S., and Pozio, S. (1996). The volcanic and tectonic history of Enceladus. *Icarus*, **119**, 385–404.

Kargel, J. S., Croft, S. K., Lunine, J. I., and Lewis, J. S. (1991). Rheological properties of ammonia–water liquids and crystal–liquid slurries: planetological applications. *Icarus*, **89**, 93–112.

Lewis, J. S. (1971). Satellites of the outer planets: their physical and chemical nature. *Icarus*, **15**, 174–85.

Lindal, G. F., Wood, G. E., Hotz, H. B., Sweetman, D. N., Eshleman, V. R., and Tyler, G. L. (1983). The atmosphere of Titan: an analysis of Voyager-1 radio occultation measurements. *Icarus*, **53**, 348–63.

Lopes-Gautier, R., Davies, A. G., Carlson, R., Smythe, W., Kamp, L., Soderblom, L., Leader, F. E., Mehlman, R., and the Galileo NIMS Team. (1997). Hot spots on Io: initial results from Galileo's near infrared mapping spectrometer. *Geophysical Research Letters*, **24**, 2439 – 42.

Lucchitta, B. K. (1980). Grooved terrain on Ganymede. *Icarus*, **44**, 481–501.

Lunine, J. I. (1989). Sulfur lakes and silicate flows: thermal emissions from Io's hot spots. *Time-variable phenomena in the Jovian system*, NASA SP-494 (ed. J. S. Belton, R. A. West, and J. Rahe), pp. 63–70. NASA, Washington, DC.

Lunine, J. I. (1989). The Urey Prize Lecture: volatile processes in the outer solar system. *Icarus*, **81**, 1–13.

Lunine, J. I., and Stevenson, D. J. (1982). Formation of the galilean satellites in a gaseous nebula. *Icarus*, **52**, 14 –39.

Lunine, J. I., and Stevenson, D. J. (1985). Physics and chemistry of sulphur lakes on Io. *Icarus*, **64**, 345– 67.

Luu, J., Marsden, B., Jewitt, D., Trujillo, C. A., Hegenrother, C. W., Chen, J., and Offutt, W. B. (1997). A new dynamical class of object in the outer solar system. *Nature (London)*, **387**, 573–5.

Masursky, H., Schaber, G. G., Soderblom, L. A., and Strom, R. G. (1979). Preliminary geological mapping of Io. *Nature (London)*, **280**, 725–33.

McCord, T. B., et al. (1997). Organics and other molecules in the surfaces of Callisto and Ganymede. *Science*, **278**, 271–3.

McCord, T. B., et al., and the Galileo NIMS Team. (1998). Salts on Europa's surface detected by Galileo's Near Infrared Mapping Spectrometer. *Science*, **280**,1242–5.

McEwen, A. S., and Soderblom, L. A. (1983). Two classes of volcanic plumes on Io. *Icarus*, **55**, 191–217.

McEwen, A. S., Lunine, J. I., and Carr, M. H. (1989). Dynamic geophysics of Io. *Time-variable phenomena in the Jovian system*, NASA SP-494 (ed. J. S. Belton, R. A. West, and J. Rahe), pp. 11–46. NASA, Washington, DC.

McEwen, A. S., Simonelli, D. P., Senske, D. R., Klaasen, K. P., Keszthelyi, L., Johnson, T. V., Geissler, P. E., Carr, M. H., and Belton, M. S. (1997). High-temperature hot spots on Io as seen by the Galileo solid state imaging (SSI) experiment. *Geophysical Research Letters*, **24**, 2443–6.

McEwen, A. S., et al. (1998). High-temperature silicate volcanism on Jupiter's moon Io. *Science*, **281**, 87–90.

McKinnon, W. B., and Melosh, H. J. (1980). Evolution of planetary lithospheres: evidence from multiringed structures on Ganymede and Callisto. *Icarus*, **44**, 454–71.

Melosh, H. J., and Schenk, P. (1993). Split comets and the origin of crater chains on Ganymede and Callisto. *Nature*, **365**, 731–3.

Moore, J. M. (1984). The tectonic and volcanic history of Dione. *Icarus*, **59**, 205–20.

Moore, J. M., and Spencer, J. R. (1990). Koyaanismuuyaw: the hypothesis of a perennially dichotomous Triton. *Geophysical Research Letters*, **17**, 1757–60.

Moore, J. M., Mellon, M. T., and Zent, A. P. (1996). Mass wasting and ground collapse in terrains of volatile-rich deposits as a solar system-wide geological process: the pre-Galileo view. *Icarus*, **122**, 63–78.

Muhleman, D. O., Grossman, A. W., Bulter, B. J., and Slade, M. A. (1990). Radar reflectivity of Titan. *Science*, **248**, 975–80.

Noll, K. S., Johnson, R. E., Lane, A. L., Domingue, D. L., and Weaver, H. A. (1996). Detection of ozone on Ganymede. *Science*, **273**, 341–3.

Noll, K. S., Roush, T. L., Cruikshank, D. P., Johnson, R. E., and Pendleton, Y. J. (1997). Detection of ozone on Saturn's satellites Rhea and Dione. *Nature (London)*, **388**, 45–7.

Pappalardo, R. T., et al. (1998). Geological evidence for solid-state convection in Europa's ice shell. *Nature (London)*, **391**, 365–7.

Parmentier, E. M., Squyres, S. W., Head, J. W., and Allison, M. L. (1982). The tectonics of Ganymede. *Nature (London)*, **295**, 290–3.

Passey, Q. R. (1983). Viscosity of the lithosphere of Enceladus. *Icarus*, **53**, 105–20.

Peale, S. J., Cassen, P., and Reynolds, R. T. (1979). Melting of Io by tidal dissipation. *Science*, **203**, 892–4.

Pieri, D. C., Baloga, S. M., Nelson, R. M., and Sagan, C. (1984). Sulphur flows of Ra Patera, Io. *Icarus*, **60**, 685–700.

Plescia, J. B. (1983). The geology of Dione. *Icarus*, **56**, 255–77.

Plescia, J. B. (1987). Geological terrains and crater frequencies on Ariel. *Nature (London)*, **327**, 201–4.

Plescia, J. B. (1988). Cratering history of Miranda: implications for geologic processes. *Icarus*, **73**, 442–61.

Plescia, J. B., and Boyce, J. M. (1982). Crater densities and geological histories of Rhea, Dione, Mimas and Tethys. *Nature (London)*, **295**, 285–90.

Plescia, B. J., and Boyce, J. M. (1983). Crater numbers and geological histories of Iapetus, Enceladus, Tethys and Hyperion. *Nature (London)*, **301**, 666–70.

Poirier, J. P. (1982). Rheology of ices: a key to the tectonics of the ice moons of Jupiter and Saturn. *Nature (London)*, **299**, 683–8.

Reynolds, R. T., Squyres, S. W., Colburn, D. S., and McKay, C. P. 1983. On the habitability of Europa. *Icarus*, **56**, 246–54.

Ross, M. N., and Schubert, G. (1985). Tidally forced viscous heating in a partially molten Io. *Icarus*, **64**, 391–400.

Ross, M. N., and Schubert, G. (1987). Tidal heating and an internal ocean model of Europa. *Nature (London)*, **325**, 133–4.

Ross, M. N., and Schubert, G. (1990). The coupled orbital and thermal evolution of Triton. *Geophysical Research Letters*, **17**, 1749–52.

Ross, M. N., Schubert, G., Spohn, T., and Gaskell, R. W. (1990). Internal structure of Io and the global distribution of its topography. *Icarus*, **85**, 309–25.

Rothery, D. A., Babbs, T. L., Harris, A. J. L., and Wooster, M. J. (1996). Colored lava flows on Earth: a warning to Io volcanologists. *Journal of Geophysical Research*, **101**, 26131–6.

Russell, M. J., and Hall, A. J. (1997). The emergence of life from iron monosulphide bubbles at a submarine hydrothermal redox and pH front. *Journal of the Geological Society*, **154**, 377–402.

Sagan, C., and Chyba, C. (1990). Triton's streaks as windblown dust. *Nature (London)*, 346, 546–8.

Schaber, G. G. (1980). The surface of Io: geologic units, morphology and tectonics. *Icarus*, **43**, 302–33.

Schenk, P. M. (1989). Crater formation and modification on the icy satellites of Uranus and Saturn: depth/diameter and central peak occurrence. *Journal of Geophysical Research*, **94**, 3813–32.

Schenk, P. M. (1991). Fluid volcanism on Miranda and Ariel: flow morphology and composition. *Journal of Geophysical Research*, **96**, 1887–906.

Schenk, P. M. (1991). Ganymede and Callisto: complex crater formation and planetary crusts. *Journal of Geophysical Research*, **96**, 15635–64.

Schenck, P. M., and Bulmer, M. H. (1998). Origin of mountains on Io by thrust faulting and large-scale mass movemements, *Science*, **279**, 544–5.

Schenk, P. M., and Jackson, M. P. A. (1993). Diapirism on Triton: a record of crustal layering and instability. *Geology*, **21**, 299–302.

Schenk, P. M., and McKinnon, W. B. (1987). Ring geometry on Ganymede and Callisto. *Icarus*, **72**, 209–34.

Schenk, P. M. and McKinnon, W. B. (1989). Fault offsets and lateral crustal movement on Europa: evidence for a mobile ice shell. *Icarus*, **79**, 75–100.

Schubert, G., Zhang, K., Kivelson, M. G., and Anderson, J. D. (1996). The magnetic field and internal structure of Ganymede, *Nature (London)*, **384**, 1514–7.

Shock, E. L., and McKinnon, W. B. (1993). Hydrothermal processing of cometary volatiles–applications to Triton. *Icarus*, **106**, 464–77.

Showman, A. P., Stevenson, D. J., and Malhotra, R. (1997). Coupled orbital and thermal evolution of Ganymede. *Icarus*, **129**, 367–83.

Simonelli, D. P., Pollack, J. B., McKay, C. P., Reynolds, P. T., and Summers, A. (1989). The carbon budget in the outer solar nebula. *Icarus*, **82**, 1–35.

Sinton, W. M., and Kaminski, C. (1988). Infrared observations of eclipses of Io, its thermophysical parameters, and the thermal radiation of the Loki volcano and environs. *Icarus*, **75**, 207–32.

Smith, B. A., et al. (1979). The Jupiter system through the eyes of Voyager 1. *Science*, **204**, 951–71.

Smith, B. A., et al. (1979). The galilean satellites and Jupiter: Voyager 2 imaging science results. *Science*, **206**, 927–50.

Smith, B. A., et al. (1981). Encounter with Saturn: Voyager 1 imaging science results. *Science*, **212**, 163–91.

Smith, B. A., et al. (1986). Voyager 2 in the Uranian system: imaging science results. *Science*, **233**, 43–64.

Smith, B. A., et al. (1989). Voyager 2 at Neptune: imaging science results. *Science*, **246**, 1422–49.

Smith, P. H., Lemmon, M. T., Lorenz, R. D., Sromovsky, L. A., Caldwell, J. J., and Allison, M. D. (1996). Titan's surface, revealed by HST imaging, *Icarus*, **119**, 336–49.

Soderblom, L. A., Kieffer, S. W., Becker, T. L., Brown, R. H., Cook, A. F., II, Hansen, C. J., Johnson, T. V., Kirk, R. L., and Shoemaker, E. M. (1990). Triton's geyser-like plumes: discovery and basic characterization. *Science*, **250**, 410 – 5.

Spencer, J. R. (1990). Nitrogen frost migration on Triton: a historical model. *Geophysical Research Letters*, **17**, 1769–72.

Spencer, J. R., and Maloney, P. R. (1984). Mobility of water ice on Callisto: evidence and implications. *Geophysical Research Letters*, **11**, 1223–6.

Spencer, J. R., Shure, M. A., Ressler, M. E., Goguen, J. D., Sinton, W. M., Toomey, D. W., Denault, A., and Westfall, J. (1990). Discovery of hotspots on Io using disk-resolved infrared imaging. *Nature (London)*, **348**, 618–21.

Squyres, S. W., Reynolds, R. T., Cassen, P. M., and Peale, S. J. (1983). The evolution of Enceladus. *Icarus*, **53**, 319–31.

Stansbury, J. A., Lunine, J. I., Porco, C. C., and McEwen, A. S. (1990). Zonally averaged thermal balance and stability models for nitrogen polar caps on Triton. *Geophysical Research Letters*, **17**, 1773–6.

Stern, S. A. (1992). The Pluto–Charon system. *Annual Reviews of Astronomy and Astrophysics*, **30**, 185–233.

Stevenson, D. J. (1982). Volcanism and igneous processes in small icy satellites. *Nature (London)*, **298**, 142–4.

Stevenson, D. J., and Lunine, J. I. (1986). Mobilization of cryogenic ice in outer solar system satellites. *Nature (London)*, **323**, 142–4.

Strom, R. G., Terrile, R. J., Masursky, H., and Hansen, C. (1979). Volcanic eruption plumes on Io. *Nature (London)*, **280**, 46–8.

Sullivan, R., et al., and the Galileo imaging team. (1998). Episodic plate separation and fracture infill on the surface of Europa. *Nature (London)*, **391**, 371–3.

Tegler, S. C., and Romanishin, W. (1998). Two distinct classes of Kuiper-belt objects. *Nature (London)*, **392**, 49–51.

Thomas, P. J., and Squyres, S. W. (1988). Relaxation of impact basins on icy satellites. *Journal of Geophysical Research*, **93**, 14919–32.

Thomas, P. J., and Squyres, S. W. (1990). Formation of crater palimpsests on Ganymede. *Journal of Geophysical Research*, **95**, 19161–74.

Trafton, L. M. (1990). Pluto's atmosphere at perihelion. *Geophysical Research Letters*, **16**, 1209–13.

Wood, J. A. (1988). Chondritic meteorites and the solar nebula. *Annual Review of Earth and Planetary Science*, **16**, 53–72.

Young, A. T. (1984). No sulphur flows on Io. *Icarus*, **58**, 197–226.

Zahnle, K., Dones, L., and Levison, H. F. (1998). Cratering rates on the galilean satellites. *Icarus*, **136**, 202–22.

Websites: images and other information are now both abundant and easy to access on the World Wide Web. I list the most useful sites below.

The *Galileo* home page covering results from *Galileo* and the *Galileo Europa Mission* is at:

http://www.jpl.nasa.gov:80/galileo/

The imaging node of NASA's planetary data system is a good way to access the archive from all NASA missions, and can be accessed at:

http://www-pdsimage.jpl.nasa.gov/PDS/

Updates on the *Cassini* mission can be found at:

http://www.jpl.nasa.gov/cassini/

The *Voyager* home page is:

http://vraptor.jpl.nasa.gov/voyager/voyager.html

The *Europa Orbiter* and *Pluto-Kuiper Express* are described at:
http://www.jpl.nasa.gov/ice_fire//

An updated list of Kuiper belt objects and related information can be found at:
http://www.ifa.hawaii.edu/faculty/jewitt/kb.html

ACKNOWLEDGMENTS

The spacecraft images used in this book all originated from NASA. Those that are not individually acknowledged below either are widely distributed standard product images or have been digitally processed by the author. The *Galileo* images reproduced here came mostly from NASA websites, and many received further digital processing by the author.

Figures

Fig. 1.8 National Space Science Data Center, World Data Center A for Rockets and Satellites (Smith, B. A.); **Fig. 2.3** modified after Consolmagno, G. J., and Lewis, J. S. (1978). *Icarus*, **34**, 284; **Fig. 2.5** modified from Schubert, G., Spohn, T., and Reynolds, R. T. (1986) in *Satellites* (ed. Burns, J. A., and Matthews, M. S.), p. 234, University of Arizona Press; **Fig. 3.2** NASA; **Figs. 3.10, 5.7c, 5.22, 5.26, 5.28, 6.1, 6.18, 6.19a, 6.23, 6.31, 6.32a, 7.2, 7.43a,** U.S. Geological Survey, Flagstaff, Arizona; **Fig. 4.3** modified from McKinnon, W. B., and Parmentier, E. M. (1986) in *Satellites* (ed. Burns, J. A., and Matthews, M. S.), p. 723, University of Arizona Press; **Fig. 4.5** modified from Stevenson, D. J. (1982). *Nature*, **298**, 144; **Fig. 5.12** National Space Science Data Center, World Data Center A for Rockets and Satellites (Kosofsky, L. J.); **Figs. 5.13, 5.16, 5.17, 6.2, 6.5, 6.7, 6.8, 6.9, 6.14, 7.7, 7.8, 7.9, 7.12, 7.26, 7.27, 7.28, 7.29, 7.30, 7.31, 7.32** *Galileo* SSI team; **Figs. 5.18, 7.13** J. R. Spencer; **Fig. 5.20** Ellsworth, K., and Schubert, G. (1983). *Icarus*, **54**, 503; **Fig. 5.23** ibid., p. 501; **Figs. 6.4, 6.15** University of London Observatory; **Fig. 6.10** C. Thomas/Lancaster University; **Fig. 6.11** modified from Shoemaker, E. H., Lucchitta, B. K., Wilhelms, D. E., Plescia, J. B., and Squyres, S. W. (1982) in *Satellites of Jupiter* (ed. Morrison, D.), p. 513, University of Arizona Press; **Fig. 6.17** modified from Moore, J. M. (1984). *Icarus*, **59**, 213; **Fig. 6.20** modified from Consolmagno, G. J. (1985). *Icarus*, **64**, 410; **Fig. 6.25** modified from Plescia, J. B. (1987). *Nature*, **327**, 202; **Fig. 6.34** modified from Smith, B. A., et al. (1986). *Science*, **233**, 62; **Fig. 6.38** modified from Janes, D. M., and Melosh, H. J. (1988). *Journal of Geophysical Research*, **93**, 3132; **Fig. 6.39** modified after Croft, S. K. (1988). *Lunar and Planetary Science Conference*, **19**, 226; **Fig. 7.14** *Galileo* NIMS Team; **Fig. 7.18** C. M. M. Oppenheimer; **Fig. 7.24** modified from Schenk, P. M., and McKinnon, W. B. (1989). *Icarus*, **79**, 80; **Fig. 7.33** J. Edmond (MIT and Woods Hole Oceanographic Institution); **Fig. 7.35** modified from Morrison, D., Owen, T., and Soderblom, L. A. (1986) in *Satellites* (ed. Burns, J. A., and Matthews, M. S.), p. 781, University of Arizona Press; **Fig. 7.37** S. A. Drury; **Fig. 7.39** modified from Croft, S. K. (1990). *Lunar and Planetary Science Conference*, **21**, 249; **Fig. 8.2** image of Titan courtesy of P. Smith and M. Lemmon (University of Arizona) and NASA; **Fig. 8.3** (top) image of Pluto courtesy of A. Stern (SwRI), M. Buie (Lowell Observatory), NASA, and ESA; **Fig. 8.3** (lower left) image of Pluto and Charon courtesy of R. Albrecht (ESA/ESO Space Telescope European Coordinating Facility), and NASA. Material in Figs 8.2 and 8.3 was created with support to Space Telescope Science Institute, operated by the Association of Universities for Research in Astronomy, Inc., from NASA contract NAS5–26555, and is reproduced here with permission from AURA/ST ScI. Any opinions, findings or conclusions in this material are solely those of its author and do not necessarily reflect the views of NASA, AURA, or ST ScI, or their employees.

Plates

Plates 3, 7 U.S. Geological Survey, Flagstaff, Arizona; **Plates 5, 6** *Galileo* solid-state imaging team.

Index

Numbers in italics refer to figures or their captions, or (where noted) to color plates. The names of specific surface features are indexed alphabetically under the name of the satellite or other body on which they occur. Note: many terms are defined in the Glossary on pages 215–22. The important physical parameters for the terrestrial planets and all satellites greater than about 200 km in radius are given in table 1.1 (page 6). The sizes of all known satellites of the outer planets are compared in figure 1.2 (page 7).